Robust Reliability in the Mechanical Sciences

Springer

Berlin
Heidelberg
New York
Barcelona
Budapest
Hong Kong
London
Milan
Paris
Santa Clara
Singapore
Tokyo

Yakov Ben-Haim

Robust Reliability in the Mechanical Sciences

With 56 Figures

Springer

Prof. Yakov Ben-Haim

Faculty of Mechanical Engineering
Technion – Israel Institute of Technology
Haifa 32000, Israel

ISBN-13: 978-3-642-64721-5 Springer-Verlag Berlin Heidelberg NewYork

Library of Congress Cataloging–in–Publication Data

Ben-Haim, Yakov, 1952-
Robust reliability in the mechanical science / Yakov Ben-Haim.
p. cm.
Includes bibliographical references and index.
ISBN-13: 978-3-642-64721-5 e-ISBN-13: 978-3-642-61154-4
DOI: 10.107/978-3-642-61154-4
1. Reliability (Engineering)—Statistical methods. 2. Robust statistics. I. Title.
TA169.B46 1996
620'.00452—dc20

Softcover reprint of the hardcover 1st edition 1996

Typesetting: Camera ready by author
SPIN: 10534776 61/3020 - 5 4 3 2 1 0 - Printed on acid-free paper

Miriam –
There is much I could never do,
Though I had all the world but you.

Preface

Reliability has a plain lexical meaning, which the engineers have modified and absorbed into their technical jargon. Lexically, that which is 'reliable' can be depended upon confidently. Applying this to machines or systems, they are 'reliable' (still avoiding technical jargon) if one is confident that they will perform their specified tasks as intended. The intuitive concept of robust reliability is that a system is reliable if it is robust with respect to uncertainty. In other words, a system is reliable if it can tolerate a large amount of uncertainty before failure can occur. Conversely, a system which is fragile with respect to uncertainty, which can tolerate only a small amount of uncertainty before failure becomes possible, is unreliable.

The analysis of reliability grapples with uncertainty. Reliability analysis begins by identifying the sources of uncertainty — loads, failure conditions, material or geometrical properties and so on — and then proceeds to the quantitative characterization of these uncertainties with a mathematical model. Various types of models of uncertainty are available — probabilistic, fuzzy, convex, etc.

In classical technical jargon, a system is reliable if the probability of failure is acceptably low. But probability is not the only viable option for quantifying uncertainty in reliability analysis, nor is it even always the most desirable mathematical model. In many complicated and important technological situations, the inability to verify the details of a probabilistic model may lead to significant inaccuracy in the reliability analysis. The idea of robust reliability developed here is an alternative to classical probabilistic reliability.

The quantification of reliability developed in this book is based on evaluation of the robustness of the system to uncertainties. The relation between noise-robustness and reliability appears widely in quality engineering, as evidenced for example by Taguchi's comment [94, p.3]:

> The broad purpose of the overall quality system is to produce a product that is *robust* [italics in the original] with respect to all noise factors. Robustness implies that the product's functional characteristics are not sensitive to variation caused by noise factors.

In our formulation, a system has high reliability when it is robust with respect to uncertainties. It has low reliability when even small amounts of uncertainty entail the possibility of failure.

The mathematical formulation of robust reliability stems from the connection between convexity and uncertainty, as developed in the theory of convex models. Robust reliability is an application of convex modelling, combined with the idea of robustness-to-uncertainty as a measure of reliability. The point of contact is the recognition of the expansion parameter of a convex model as the uncertainty parameter whose limiting magnitude, at failure of the system, assesses the uncertainty-robustness of the system.

Convex-set-models of uncertainty, rather than probabilistic models, have a quite diversified history in the engineering literature of the past few decades, representing complex uncertainties in a range of technological applications. We mention here just a few examples. Drenick [33, 34] and Shinozuka [84] describe uncertain seismic loads on civil structures by defining convex sets of possible input functions, with no probability measures defined on these sets. Schweppe [80] and Witsenhausen [104, 105] describe estimation and control algorithms for linear dynamic systems based on sets of inputs. Schweppe [81] develops inference and decision rules based on assuming that the uncertain phenomenon can be quantified in such a way as to be bounded by an ellipsoid, again with no probability function involved. Ben-Haim [6] develops a method for optimal design of material assay systems based on convex sets of uncertain spatial distributions of the analyte material. Ben-Haim and Elishakoff [20] describe a range of analysis and design problems in applied mechanics based on defining convex sets of uncertain input functions or uncertain geometrical imperfections. Lindberg [61, 62] and Ben-Haim [11] use the convex modelling method to study radial pulse buckling of geometrically imperfect thin-walled shells. Elishakoff and Zhu [38] use convex models to represent uncertainties in the acoustic excitation of structures. Natke and Soong [73] study the topological optimization of mechanical structures with convex model representation of uncertain dynamic loads on the structure.

The modern theory of random vibration of structures has developed over the past few decades in parallel to, and quite independently of, convex models of uncertainty. Much current reliability theory for mechanical systems and structures is based on the probabilistic theory of vibrations. A plethora of excellent monographs and textbooks are devoted to random vibrations, and many discuss implications for reliability. However, the systematic inclusion of probabilistic reliability in university programs in mechanical, civil, aerospace and manufacturing engineering has not followed the surge of books on random vibrations. The reasons for this are undoubtedly varied, and any author on *non-probabilistic* reliability who offers an explanation is inviting unmitigated criticism. Nonetheless, one factor underlying the reticence of engineering curricula in the area of probabilistic reliability is the problem of applicability. In many practical situations one lacks the information needed to verify the probabilistic models upon which probabilistic reliability analysis is based.

This is discussed in chapter 7 and elsewhere [14, 16, 20]. A major aim of this book is to develop a methodology for reliability analysis which is particularly suited to the type of partial information characteristic of mechanical systems and structures.

That reliability has not widely reached university curricula in mechanical sciences is no reflection on the practical importance of the subject. What has happened in many institutions is that the academic responsibility for subjects such as reliability, quality control, monitoring and fault detection, has been transferred from the technological departments to departments of industrial management or business administration. These subjects are important in managerial training, but will inevitably suffer from lack of technological expertise. When a department within the domain of mechanical sciences teaches reliability, it can responsibly address the technical questions of design and analysis of reliable systems. These subjects are essential complements to the generic management approach to reliability which is accessible to non-engineering students. The present book is directed to engineering students, and provides an implementable application of their technological training.

The main emphasis in this book on reliability in mechanical science is on mechanical devices and structures. That is, vibrating solid constructions subject to external loads and internal imperfections, such as aerospace structures, buildings, bridges, machine tools, and mechanical devices of all sorts. In addition, the reliability analysis of fluid flow systems such as turbo-machinery, of heat-transfer devices such as cooling fins, and of manufacturing processes such as extrusion and milling, are explored in examples and homework problems.

•

The literature on convex modelling is technical and specialized. The book before you is the first which deals with an application of convex models on a level accessible to students and practitioners of engineering who are not research specialists. Designed as an upper-level undergraduate or first-year graduate text on robust reliability of mechanical systems, this book presumes a basic knowledge of strength of materials and vibration dynamics, linear algebra and integral and differential calculus. Chapters 7 and 8 also assume familiarity with the basic concepts of probability. The aim of the book is to give the student or engineer a working knowledge of robust reliability, which will enable him to analyze the reliability of mechanical systems. Each chapter is introduced with a brief conceptual survey of the main ideas, which are then developed through examples. Problems at the end of each chapter give the student the opportunity to strengthen and extend his understanding. Problems marked with an double dagger (‡) are more difficult or open-ended.

The book is divided into 9 chapters.

Chapter 1 is a brief qualitative preview of the idea of robust reliability. We discuss some practical examples, without going into the mathematical tools of reliability analysis.

Representation of uncertainty by convex models, rather than by probability density functions, provides the mathematical basis of robust reliability. Selected geometrical and algebraic properties of convex models are summarized in chapter 2. With the exception of chapters 7 and 8, probabilistic ideas are not employed in this book.

Chapters 3 and 4 constitute the heart of our presentation of robust reliability. In these chapters we develop the method of robust reliability based on convex models, first with static and then with dynamic systems. Subsequent chapters address various related subjects, many of which are analogous to concepts arising in the classical probabilistic theory of reliability.

Chapter 5 deals with fault diagnosis, system identification and reliability testing. In the context of robust reliability this means that we seek to identify the system and associated uncertainties, in order to determine if the performance will be robust with respect to the estimated degree of uncertainty. In considering fault diagnosis, we emphasize assessment of the robust reliability of the diagnostic algorithm itself. Since the full range of tools of parameter estimation and system identification are beyond the scope of this book, we develop only those methods which are inherent in the analysis of robust reliability with convex models of uncertainty.

Robust reliability, as developed in the previous chapters, is a property of the mechanical system and its operating environment. However, the same analysis can be applied to evaluating both the reliability of mathematical models of systems, and the robustness of design or operational decisions based on these models. Chapter 6 introduces the analysis of robust reliability of mathematical models of mechanical systems.

While this book does not develop a probabilistic theory of reliability, we do study the relation between probabilistic and robust reliability from various points of view. In chapter 7 we develop the idea that uncertainty and probability are not synonymous; other non-probabilistic mathematical models can be used to quantify the phenomena of uncertainty. Following this, we discuss some limitations of probabilistic theory for reliability analysis.

Despite the considerations discussed in chapter 7, there are situations in which verified probabilistic models are available to the mechanical designer or analyst. The stochastic Poisson process is one case in point. In some situations, information about the distribution of discrete events in time or space can be confidently modelled statistically using the Poisson distribution. In chapter 8 we develop a hybrid robust-probabilistic analysis of reliability.

Chapter 9 concludes the book with a brief summary and discussion of some speculative aspects of reliability. Most importantly, we consider the problem of "calibrating" a reliability analysis, and establishing subjective interpretations of a quantitative measure of reliability.

"The writing of books", said Solomon,[1] "has no end", but also their beginnings are often indiscernible, rising unsuspectedly from innumerable chats, discussions, arguments and correspondences. And so it was with this book. I cannot hope to trace the sources of the thoughts which come together here. I can however, with great pleasure, acknowledge a few of my friends and colleagues who, by untiring argument, relieved me of some of my worst ideas and greatest mistakes: Prof. Gerard Lallement and Dr. Scott Cogan from Université de Franche-Comté in Besançon, France, as well as Prof. H.G. Natke and Dr. Uwe Prells from Universität Hannover in Germany. The errors which remain are entirely my own.

In addition I am both proud and happy to acknowledge the unsurpassed atmosphere of free thought and open enquiry at the Technion—Israel Institute of Technology.

Last but not least, thousands of unneeded words have been avoided by the excellent graphic art of Ms. Ariela Rozen. I am sure the reader will join me in expressing heartfelt thanks for her efforts!

[1] *Kohelet* (Ecclesiastes) 12:12.

Contents

Chapter 1

Preview of Robust Reliability

Technological systems are designed to perform well-defined tasks. However, these systems and their environments are often inordinately complex, and the designer invariably suffers from incomplete knowledge of the properties of the system and its environmental conditions. One aim of reliability analysis and design is to enhance the robustness of the system to the uncertainties inherent in limited information. A system is reliable if it is robust with respect to these uncertainties. In other words, a system is reliable if it will perform satisfactorily in the presence of large uncertainties. On the other hand, a system is unreliable if it can fail due to even small deviations from nominal circumstances.

In this book we will develop the tools of robust reliability analysis. This involves a modicum of mathematics which, though elementary, might be unfamiliar to the reader. We defer the mathematics, however, to subsequent chapters. In this chapter we will discuss several simple qualitative examples of robust reliability, without any mathematics. This preview will equip the reader with an intuitive feeling for the subject, and prepare him for the quantitative work ahead.

The analysis of the reliability of a mechanical system will invariably employ three components: (1) a model of the mechanical properties and physical laws which govern the system, (2) conditions for failure of the system, and (3) a model of the uncertainties which accompany the system. These uncertainties may be in the operational environment as well as in the mechanical model and failure criteria. The reliability is assessed as the greatest amount of uncertainty consistent with successful operation of the system. The uncertainty is represented with convex models, which are discussed in chapter 2, so for the duration of this preview chapter we will be somewhat vague about the "uncertainty parameters" which measures the "amount" of uncertainty.

Our aim is to get a feel for robust reliability analysis, without worrying about how the analysis is actually accomplished.

1.1 Flexible Solar Panels

Many contemporary satellites carry panels of solar collectors borne by lightweight flexible truss structures. These panels deploy in space by folding or snapping open. Subsequently, they must maintain a stable orientation with respect to the body of the satellite.

A possible cause of mechanical failure of the solar panel is the emergence of large-amplitude vibrations. The truss could suffer damage or even complete destruction if these vibrations exceed specific limits. These vibrations result from dynamical coupling of the truss to the satellite body, which performs gradual maneuvers and also displays mechanical vibrations.

Quite a lot may be known about the forces on the truss resulting from planned satellite maneuvers. Speed and acceleration schedules for standard maneuvers may be known. There may also be information about the mechanical vibrations of the satellite body. The frequency range of these vibrations as well as something about their magnitudes may be known from ground tests. However, there nonetheless remains considerable uncertainty about the precise waveforms of these complicated vibrational excitations.

A further important uncertainty relevant to the solar truss is the imprecise knowledge of the mechanical properties of the truss itself. In particular, damping and stiffness properties of the joints and beam-elements may be imperfectly known.

Failure-prevention by suppression of vibrations in the truss must be achieved by passive damping and/or active feedback control. In both cases, the task is complicated by the lack of information about the excitations and about the truss itself.

We have now described in qualitative terms the three components essential for a reliability analysis: the mechanical properties of the system, the conditions for failure, and the uncertainties which accompany the system. The system is reliable if it is robust with respect to the uncertainties. The reliability is measured by the maximum amount of uncertainty which is consistent with successful performance of the intended mission of the system.

In reliability analysis we attempt to identify those factors which control the robustness to uncertainty. This analysis has immediate design implications, by suggesting design decisions which enhance the reliability. Speaking a bit more formally, we will identify an "uncertainty parameter" α which quantifies the amount of uncertainty. (Sometimes there is more than one uncertainty parameter). The crux of the reliability analysis is the determination of the greatest value of α which is consistent with no-failure. The maximum is denoted $\widehat{\alpha}$, and is a function of the properties of the system, in particular, it depends on the design variables which are available for modification. The design values are chosen to maximize $\widehat{\alpha}$: to make the system resistent to uncertainty to the greatest possible extent.

1.2 Quality Control of Thin Shells

Thin-walled shells are widely used structural elements, and display favorable weight-to-strength ratios. However, the strength may be drastically diminished by even small geometrical imperfections in the shape of the shell. Geometrical imperfections are unavoidable in the manufacturing process, so real shells will deviate in an unknown manner from the ideal design shape. The disparity between real and ideal shell is inevitable, but its magnitude is unknown, and this information gap between design and implementation is the starting point for robust reliability analysis. The central question is: how large an uncertainty in the shell shape can be tolerated? A reliable shell-design will not fail under load even with large geometrical uncertainty. An unreliable design can fail even with minute shape imperfections.

In this example we meet a confluence of reliability analysis and quality control. From the quality-control point of view, the maximum acceptable geometric uncertainty in the shell is precisely the radial tolerance to which the shell should be manufactured. In other words, the output of the reliability analysis (the greatest tolerable amount of uncertainty) is the input to the quality control (the radial tolerance). Furthermore, at the reliability stage we can inquire about the spatial variation of the acceptable uncertainty over the surface of the shell. In this way we obtain a radial tolerance which varies from point to point on the shell surface.

We have considered only the amplitude of geometrical imperfections, but we can also consider the area subtended by an imperfection. Again, the reliability analysis will indicate the maximum acceptable area of an imperfection. Both the amplitude and the area of the geometric imperfection influence the performance of the shell. As one aspect of the uncertainty is allowed to increase the other must be constrained if one wishes to maintain the shell strength. This has implications for inspection and quality-certification, since it indicates the measurement sensitivity and spatial sampling rate required for safety certification.

In this example we have two uncertainty parameters: one for the amplitude of the geometrical imperfection and one for its area. For a fixed area α_{ar} of imperfection, let $\widehat{\alpha}_{\mathrm{am}}(\alpha_{\mathrm{ar}})$ denote the greatest acceptable imperfection amplitude. We would expect that an imperfection which subtends a small area can achieve greater amplitude without allowing failure, than an imperfection of large area. In other words, $\widehat{\alpha}_{\mathrm{am}}$ versus α_{ar} will usually be a decreasing curve, as in fig. 1.1. This "reliability curve" divides the plane into a "safe" and an "unsafe" region. Any point $(\alpha_{\mathrm{am}}, \alpha_{\mathrm{ar}})$ below the curve corresponds to acceptable magnitudes of uncertainty, both in area and amplitude. Any point above the curve represents unacceptable uncertainty, and entails the

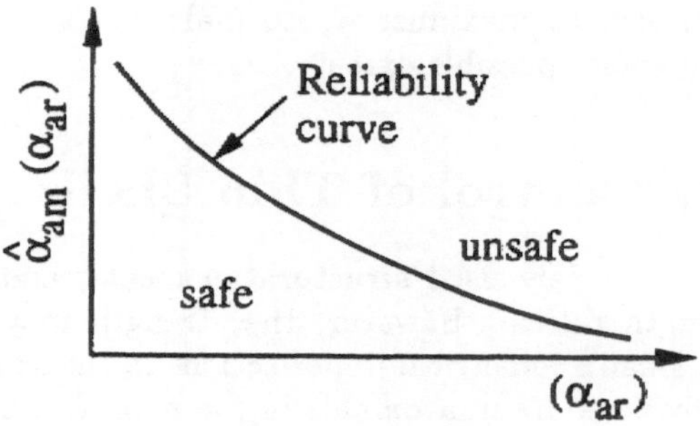

Figure 1.1: Schematic reliability curve for geometric imperfections of shells.

possibility of failure of the shell under design loads.

1.3 Fatigue Failure and Reliability

Solid materials subject to low-amplitude repetitive loading ultimately fail due to fatigue cracking. This phenomenon of 'low-cycle fatigue' has been extensively studied in the laboratory environment with single-frequency harmonic load cycles applied to small test pieces. The standard laboratory assessment of fatigue lifetime of a material is the S–N curve: the amplitude S of the harmonic load versus the number N of cycles to failure. This curve depends on the choice of material and on the mean load level.

For a solid structure excited by a single-frequency harmonic load, one could calculate the number of cycles to fatigue failure, based on a model of the structural vibration dynamics and the S–N curves of the material from which the structure is made. In practice however load cycles are hardly ever composed of only a single frequency. Rather, realistic load histories are complicated and variable, and the S–N-calculation is likely to be inaccurate. Extensive evidence indicates that the net damage from a sequence of high- and low-amplitude load cycles is quite sensitive to the order in which these cycles are applied.

The reliability questions which arise are: how sensitive is the structure to load-uncertainty? What design properties of the structure are dominant in determining this sensitivity? What classes of uncertain loads are particularly pernicious in reducing the lifetime? The robust-reliability approach to these questions is to determine the maximum tolerable uncertainty, $\widehat{\alpha}$: the greatest value of the load uncertainty parameter for which the structure will not fail earlier than a specified time. The structure is reliable is $\widehat{\alpha}$ is large, while if

$\widehat{\alpha}$ is small then the structure is liable to fail early in life if small deviations of the load history occur. $\widehat{\alpha}$ depends on the properties of the system and the class of uncertain inputs. In comparing two alternative structures or designs, their $\widehat{\alpha}$-values express their comparative reliability.

1.4 Plastic Extrusion Manufacturing

In most of our work we will be concerned with the reliability of physical systems: structures, mechanical devices, and so on. However, the same concept of robust reliability can be applied to mathematical models of systems. We consider here a simple example from the field of manufacturing technology.

A small company produces plastic parts by high-temperature extrusion molding. The manufacturing process is characterized by several dynamic control variables: pressures, temperatures and flow rates at different points in the system. We will collectively denote these control variables by a vector x. The control variables are liable to fluctuate, but the system is flexible, so that drift of one variable can be compensated by alteration of another. The status of the manufacturing process is assessed by an empirical performance function $f(x)$, which is a polynomial in the control variables. The process is acceptable provided that the performance function is zero: $f(x) = 0$. A control system modulates the control variables to maintain this condition.

However, the performance polynomial $f(x)$ is not perfect: its coefficients are based on past experience and thus are uncertain. How reliable is the manufacturing procedure, when operated to preserve $f(x) = 0$? How much uncertainty in the coefficients of the performance function can the process tolerate? The process is reliable with respect to the uncertainty in the model if the manufactured product is acceptable even when the coefficients are quite wrong. On the other hand, the system is unreliable if it is fragile with respect to uncertainty in the coefficients; it is unreliable if small fluctuations in the model can result in unacceptable performance. Furthermore, we anticipate that the robustness may be different for different coefficients of $f(x)$. While we will define an "overall reliability", we will also be able to evaluate specific "conditional reliabilities" for individual coefficients. These conditional reliabilities define a "reliability surface" which is a multi-dimensional generalization of fig. 1.1.

The reliability analysis of a mathematical model has three components, just like the analysis of system-reliability described on page 2. The components are (1′) a decision algorithm based on the mathematical model of the system, (2′) a criterion for failure of the decision algorithm and (3′) a model of uncertainty of the mathematical model, as well as other relevant uncertainties.

In the plastic extrusion example, the control algorithm "makes decisions" and issues commands based on the performance function, which is the mathematical model in question. The controller fails if the product is unsatisfactory,

Example	Model	Failure Criterion	Uncertainty
Solar Panel	Truss	Large amplitude vibration	Excitation, damping
Shell	Buckling mechanics	Buckling load	Shell shape
Fatigue	Vibration dynamics	Energy dissipation	Excitation
Extrusion	Performance function	Product acceptability	Coeffs. of $f(x)$

Table 1.1: Recapitulation of examples.

and the uncertainty lies in the coefficients of the performance function.

1.5 Summary

Table 1.1 recapitulates the examples of robust reliability discussed in the previous four sections. In this book, reliability is assessed in terms of robustness to uncertain variation. The system or model is reliable if failure is avoided even when large deviations from nominal conditions occur. On the other hand, a system is not reliable if small fluctuations can lead to unacceptable performance. Robustness and fragility describe opposite extremes of reliability.

In the solar panel example, the mechanical properties can be expressed by a truss model, while failure occurs if the vibration amplitude exceeds a given limit. The uncertainties are in the excitations as well as in the damping properties of the truss. The solar collector is reliable if unacceptable vibration amplitudes cannot occur even if the uncertainties are large. On the other hand, for example, the panel is unreliable if small deviations in the damping properties from the design specifications can result in unacceptably large oscillations.

The thin-walled shell is described by a mechanical buckling model, and it fails if the buckling load of the shell is exceeded. The buckling load is sensitive to geometrical imperfections, which are uncertain. The shell is unreliable if small shape imperfections entail the possibility of failure.

Fatigue failure can occur in a structure subjected to variable low-amplitude cyclic loading. The behavior of the structure is described by a vibration dynamics model. The evolution of damage is related to the dissipation of energy in the structure which eventually leads to fatigue failure. In practical situations, the load patterns are variable and imprecisely known. The structure is reliable if it will avoid failure for a specified duration even in the presence of large load uncertainty. On the other hand, the structure is unreliable if small deviations from design-loads can lead to premature failure.

In plastic extrusion, an empirical performance function is used to regulate plant variables. The coefficients of the function are uncertain, but the manufacturing process is reliable if the product is acceptable even with large parameter uncertainty.

Chapter 2

Convexity and Uncertainty

If you ask a person, 'what do you know?' he can tell you. If you ask him, 'what do you *not* know?' what can he say? The phenomena of uncertainty lie in that tantalizing gap between what we *do* know and what we *could* know.

2.1 Complex Uncertainty and Limited Information: Four Examples

Every quantitative theory of reliability takes as its starting point a mathematical model for representing and quantifying uncertainty. The concept of robust reliability developed in this book is based on convex models of uncertainty. In this chapter we describe the basic properties of convex models which will be needed later on. Convex models are particularly suited for modelling the complex uncertain events which arise in many mechanical and structural applications. We begin with four brief case studies of uncertainty represented by convex models [16].

A convex model is a convex set of functions or vectors. Each element of the set represents a possible realization of an uncertain event. The set as a whole expresses the amount of variation of these events. In the examples that follow, these events are spatial distributions of imperfections in a structure, or temporal variation of the ground motion during an earthquake, and so on. The convex model, by its structure and position in the space within which it is defined, reflects what is known about the events: what types of events occur. On the other hand, the size of the set reflects the variability of the events: the gap between what we expect and what could actually transpire.

Buckling of thin-walled shells. Thin-walled shells such as cylinders and domes have very high load-bearing capacities compared to their weight.

For example, a sheet of paper stood on end buckles under slight pressure, while if rolled into a cylinder and taped it can withstand an axial loading of considerable weight. However, small geometrical imperfections in the shape of the shell can drastically reduce the maximum load which the shell can carry.

A typical engineering question which arises is: what radial tolerance in the shell-shape assures that the weakest shell will suffer a reduction in buckling-load[1] by no more than a specified amount? In a more sophisticated analysis one recognizes that the boundary conditions of the shell, together with the shell dynamics, can allow greater tolerance in some regions of the shell than others. Consequently, one can ask: what variation of the shell-shape tolerance over the surface of the shell is allowed?

The first problem one confronts in addressing these questions is: how to model the range of possible shell shapes coming off a production line? Some information is available from actual measurements of shell shapes, though it is scanty and very expensive (see, for example, [2, 53]). However, one can readily formulate an infinite set of functions which represents the uncertainty of the shapes: each function represents a particular shell shape, while the set expresses the uncertainty in which shape will actually occur. This is precisely a set-model for shell-shape uncertainty; when the set is convex, it is a convex model. Furthermore, one can do so in such a way that the set depends parametrically on the radial tolerance of the shell. Then it is possible to evaluate the buckling load of the weakest shell as a function of the radial tolerance [19, 20, 37]. In this way, the convex model incorporates uncertainty in the shell shape into the design and manufacturing procedure, without relying on probabilistic information.

Vehicle vibrations on rough terrain. A vehicle traversing rough terrain can induce discomfort or even functional incapacity in the passengers as well as damage to on-board equipment. A reliable suspension system is optimized to reduce these effects. The optimization must be performed with respect to a model of the uncertain terrain. Statistical models have been employed for representing the variation of uncertain terrain [74]. However, verification of these models is time consuming and expensive. Alternatively, global features of the surface which are comparatively easily measured, such as maximum roughness or slope variation or other features, can be used to define sets of possible substrates. These sets are convex models of the substrate uncertainty. The design decisions are then made so as to assure that the worst ride (e.g. maximum instantaneous acceleration) induced by any allowed substrate is acceptable [20, 21].

Seismic safety. A major challenge in civil engineering design for seismically active regions is the prevention of life-threatening structural damage resulting from earthquakes. Also important is the amelioration of seismically-induced damage to equipment and secondary systems (power systems, com-

[1] The buckling load is the least load at which the shell fails by buckling.

munications equipment, etc.). The design of seismically reliable structures is complicated by the wide temporal and spatial variability of ground motion during an earthquake and its complex interaction with structures.

Global features of seismic events, such as total or instantaneous energy, can be used to define sets of seismic events which include extreme cases more explicitly than probabilistic models. It is fairly straightforward to formulate a convex model as the set of all seismic events consistent with available fragmentary data constraining the seismic occurrence. One can then optimize the design with respect to this set of conceivable events. In this way information about earthquake-uncertainty is incorporated in the structural design and reliability analysis without postulating stochastic properties of seismic phenomena [33, 34, 84].

Design of a reliable pressure vessel. A standard task in structural engineering design is to choose the wall thickness of a pressure vessel subject to uncertain internal fluid pressure. If the probability density function of the pressure fluctuations is known, then it is a matter of fairly straightforward analysis to determine the least wall thickness needed to assure that the probability of failure by yielding is less than a specified amount. However, if high reliability is required (low probability for failure), then even very small errors in the tails of the probability density can result in large errors in the chosen wall thickness [14; 20, pp.11–13]. A hybrid probabilistic–non-probabilistic approach is possible here. One defines the set of all probability densities which are consistent with available information. This set is a convex model for the uncertainty in the probability density. The wall thickness is then chosen with respect to this non-probabilistic specification of the uncertainty in the probability density of the pressure.

2.2 Some Convex Models

In this section we define a range of convex models of uncertainty, and discuss typical engineering considerations underlying the selection of a model.

Energy-bound models. Consider a warped beam which deviates by $y(x)$ from its nominal shape at position x along the length of the beam. Energy is required to straighten out such a beam. Or, conversely, energy was required to produce the warping. Specifying the amount of energy required still leaves some uncertainty as to the original shape of the beam. One type of energy-bound convex model of the shape-uncertainty is defined as the set of all beam-profiles requiring no more than an amount α of elastic energy to straighten them out [12]. This set of profiles is one type of convex model:

$$\mathcal{Y}(\alpha) = \left\{ y(x) : \ \frac{EI}{2} \int_0^L (\ddot{y}(x))^2 \, dx \ \leq \ \alpha \right\} \tag{2.1}$$

where L and EI are the length and flexural rigidity of the beam, respectively, and dots imply differentiation with respect to position. (It is implicitly assumed that the elements of $\mathcal{Y}(\alpha)$ satisfy the boundary conditions inherent in the mechanical system.)

Energy-bounds can be related to uncertainty in many situations. In section 2.1 we mentioned energy-bound models representing uncertainty of seismic input waveforms, in terms of bounds on the total or instantaneous ground-motion energy. Not infrequently the 'energy' is loosely defined and, in analogy to the energy of an electric current, the convex model is defined as a bound on a quadratic function. For example, if $u(t)$ is a scalar function representing the uncertain ground motion as a function of time, a common energy-bound uncertainty model is:

$$\mathcal{U}(\alpha) = \left\{ u(t) : \int_0^T u^2(t)\,\mathrm{d}t \le \alpha \right\} \tag{2.2}$$

Energy-bound models can be defined for vector functions as well:

$$\mathcal{U}(\alpha) = \left\{ u(t) : \int_0^T u^T(t) V u(t)\,\mathrm{d}t \le \alpha \right\} \tag{2.3}$$

where V is a positive definite real symmetric matrix. The convex models of eqs.(2.2) and (2.3) are often called "cumulative energy-bound models", to distinguish them from the "instantaneous energy-bound model" which is the set of functions whose "energy" is constrained at each instant:

$$\mathcal{U}(\alpha) = \left\{ u(t) : u^T(t) V u(t) \le \alpha \right\} \tag{2.4}$$

An enormously popular 'energy-bound' uncertainty model is the ellipsoid-bound model for vectors, studied extensively by Schweppe [81]. If v is the uncertain vector, then an ellipsoidal model for the uncertainty in v is the set of vectors contained within an ellipsoid:

$$\mathcal{V}(\alpha) = \left\{ v : v^T W v \le \alpha \right\} \tag{2.5}$$

where W is a real symmetric positive definite matrix.

An immediate variation on the ellipsoid-bound model is the shifted ellipsoid model, where the ellipsoid is centered around a nominal or reference point, $\overline{v}$:

$$\mathcal{V}_S(\alpha) = \left\{ v : (v - \overline{v})^T W (v - \overline{v}) \le \alpha \right\} \tag{2.6}$$

Since $\mathcal{V}_S(\alpha)$ is simply a shifted version of $\mathcal{V}(\alpha)$, these two convex models are related as:

$$\mathcal{V}_S(\alpha) = \mathcal{V}(\alpha) + \overline{v} \tag{2.7}$$

where the + operator means that $\overline{v}$ is added to each element of $V(\alpha)$.

Envelope-bound models. The need to represent geometric uncertainties, as well as other applications, gives rise to envelope-bound convex models.

Forces acting on an unknown domain of a structure [8], or obstacles of unknown size and position in air ducts [9], or local imperfections in shells [11] are all amenable to representation by envelope-bound models. Let $g(x)$ be the uncertain function of a spatial variable x. An envelope-bound convex model is:

$$\mathcal{E}(g_1, g_2) = \{g(x) : \ g_1(x) \le g(x) \le g_2(x)\} \tag{2.8}$$

where $g_1(x)$ and $g_2(x)$ define the bounding envelope. To take a specific example, consider a beam of length L, so $0 \le x \le L$. Suppose the beam is warped in the interval $[a, b]$ but otherwise straight, and let $g(x)$ represent the imperfection profile of the beam. By choosing the envelope functions g_1 and g_2 as follows, $\mathcal{E}(g_1, g_2)$ can represent uncertain local damage of magnitude not exceeding α:

$$g_n(x) = \begin{cases} 0 & x \notin [a, b] \\ (-1)^n \alpha & x \in [a, b] \end{cases}, \qquad n = 1, 2 \tag{2.9}$$

Minkowski-norm models. The quadratic term $v^T W v$ in the convex model of eq.(2.5) is the square of a vector norm. This *weighted euclidean norm*, $\sqrt{v^T W v}$, is a generalization of the euclidean length of the vector v, which results when W is the identity matrix. But the weighted euclidean norm is itself a special case of the *Minkowski norm*, which is useful in defining a wide class of convex models. For $r \ge 1$, the Minkowski norm of a vector $x \in \Re^N$ is:

$$||x||_r = \left(\sum_{n=1}^{N} |x_n|^r \right)^{1/r} \tag{2.10}$$

When $r = 2$ and $x = W^{1/2} v$, we obtain the weighted euclidean norm:

$$||x||_2^2 = ||W^{1/2} v||_2^2 = v^T W v \tag{2.11}$$

A Minkowski-norm convex model is defined as:

$$\mathcal{W}_r(\alpha) = \{x : \ ||x||_r \le \alpha\} \tag{2.12}$$

A remarkable thing happens to the Minkowski norm $||x||_r$ as r tends to infinity [48, p.15]:

$$\lim_{r \to \infty} ||x||_r = \max_n |x_n| \tag{2.13}$$

This property of the Minkowski norm allows us to formulate an envelope-bound convex model. The symmetric envelope-bound convex model for vectors $x \in \Re^N$ contains all vectors whose elements are bounded by given values:

$$\mathcal{E}(\widehat{x}) = \{x : \ |x_n| \le \widehat{x}_n, \quad n = 1, \ldots, N\} \tag{2.14}$$

To represent this convex model with the Minkowski norm, define a diagonal matrix $\widehat{X} = \mathrm{diag}(1/\widehat{x}_1, \ldots, 1/\widehat{x}_N)$. The symmetric envelope-bound convex model $\mathcal{E}(\widehat{x})$ can be represented as follows.

$$\mathcal{W}_\infty(\widehat{X}) = \left\{x : \ \|\widehat{X}x\|_\infty \le 1\right\} \tag{2.15}$$

In its most general form, the bounds of the envelope-bound convex model are not symmetric with respect to zero. Such a set is simply a translation of $\mathcal{W}_\infty(\widehat{X})$.

Slope-bound models. The envelope-bound concept can be applied to the slope rather than to the magnitude of a spatially uncertain quantity. Such convex models have been used in analysis of vehicle dynamics on barriers and uncertain rolling terrain [21]. Similarly, in modelling uncertain heating processes [18] the uncertain function may be constrained to increase monotonically between given limits, but to be otherwise of unknown variation.

For example, let $r(t)$ be the heat flux out of a nuclear reactor fuel element which, during a transient, increases monotonically between r_1 and r_2 over the time interval $[0, T]$. A slope-bound convex model for the uncertainty in $r(t)$ during the transient is:

$$\mathcal{R} = \left\{r(t) : \ r(0) = r_1, \ \ r(T) = r_2, \ \ \frac{dr}{dt} \ge 0\right\} \tag{2.16}$$

Fourier-bound models. In many situations the engineer has partial spectral information for characterizing an uncertain phenomenon. For example, geometric shape-imperfections of thin walled shells, mentioned in section 2.1, have been measured spectrally [2, 53]. Data such as these lead to ellipsoid-bound models for the uncertainty in the spectral coefficients. Let c represent a vector of Fourier coefficients of the shape of the geometric imperfection. A Fourier ellipsoid-bound convex model of uncertainty is [11, 37, 61, 62]:

$$\mathcal{C}(\alpha) = \left\{c : \ (c - \overline{c})^T \, W \, (c - \overline{c}) \le \alpha^2\right\} \tag{2.17}$$

where $\overline{c}$ is a nominal Fourier-coefficient vector, W is a positive definite real symmetric matrix determining the shape of the ellipsoid and α determines the size of the ellipsoid and the amount of uncertainty. $\mathcal{C}(\alpha)$ is in fact an energy-bound convex model, like eq.(2.6).

Spectral envelope-bound models are also used. If $u(\omega)$ is an uncertain Fourier transform, then a Fourier envelope-bound model is [14]:

$$\mathcal{U}(u_1, u_2) = \{u(\omega) : \ u_1(\omega) \le |u(\omega)| \le u_2(\omega)\} \tag{2.18}$$

where $|u(\omega)|$ is the absolute value of the complex function $u(\omega)$, and $u_1(\omega)$ and $u_2(\omega)$ are real envelope functions.

Mass distribution models. In the assay of material it sometimes occurs that very little is known about the possible spatial distributions which the material can assume. Such problems arise in nuclear radiological measurements [95], in nuclear waste assay [6, 24, 83], in subterranean geological prospecting [101] and elsewhere. The simplest convex model for representing unknown spatial distributions of material is the distribution-function model. Let $m(x)$ be the density of analyte material at position x, distributed over domain X. The set of allowed distributions of total mass μ_0 is:

$$\mathcal{M}_0(\mu_0) = \left\{ m(x) : \ m(x) \geq 0, \quad \int_X m(x)\, \mathrm{d}x = \mu_0 \right\} \tag{2.19}$$

In some situations, the information which constrains the allowed spatial distributions of analyte material is the nth (usually 1st or 2nd) moments of the spatial distribution. In this case, a convex model for uncertainty in $m(x)$ is:

$$\mathcal{M}_n(\mu_n) = \left\{ m(x) : \ m(x) \geq 0, \quad \int_X x^n m(x)\, \mathrm{d}x = \mu_n \right\} \tag{2.20}$$

2.3 Expansion of Convex Models

A convex model is a set which, when thought of geometrically, has a particular shape and size. The shape of the set indicates how the uncertain events cluster, and the size of the set tells something about how much variation or uncertainty is anticipated. Consider for example an N-vector function $u(t)$ representing a time-varying load distributed at N points on a structure. The nominal load history vector is $\overline{u}(t)$, but actual load histories may deviate from $\overline{u}(t)$. If we know an upper bound on the "instantaneous energy" of the load deviation from $\overline{u}(t)$, then we could use the instantaneous energy-bound convex model to represent the uncertainty in the load vector:

$$\mathcal{U}(\alpha) = \left\{ u(t) : \ [u(t) - \overline{u}(t)]^T [u(t) - \overline{u}(t)] \leq \alpha^2 \right\} \tag{2.21}$$

This is a set of N-dimensional functions. What is the "shape" of this set? A solid sphere in N-dimensional euclidean space, centered at the point $\overline{x} = (\overline{x}_1, \ldots, \overline{x}_N)$ and of radius R, is the set of points:

$$\mathcal{X}(\alpha) = \left\{ x : \ \sum_{n=1}^{N} (x_n - \overline{x}_n)^2 \leq R^2 \right\} \tag{2.22}$$

Or, equivalently, the solid sphere is the set of points $x \in \Re^N$:

$$\mathcal{X}(\alpha) = \left\{ x : \ (x - \overline{x})^T (x - \overline{x}) \leq R^2 \right\} \tag{2.23}$$

By analogy to eq.(2.23) we can think of the convex model $\mathcal{U}(\alpha)$ as a solid sphere of radius α in the space of functions.

The "size" of the set $\mathcal{U}(\alpha)$ is α, in analogy to the radius R of the sphere. There are other ways of indicating size, such as diameter, circumference, volume and so on. So when we refer to the size of a convex model we do not specify all there is to know about the set, but only give an indication, quantitative but partial, of the range of variation of the elements of the convex model.

The shape of a convex model is determined by the "guiding principle" by which the uncertain events cluster, while its size tells us how "tightly" the events cluster. In using a convex set to represent uncertainty we refer to the size parameter, such as α in eq.(2.21), as a *measure of uncertainty*. In the geometry of convex sets the size parameter is called the *expansion parameter* of the set. The set $\mathcal{U}(\alpha)$ expands and contracts like a balloon as α grows and diminishes, retaining its shape but varying its dimensions. In robust reliability analysis we will often ask: how much can the convex model of uncertainty expand before failure becomes possible. The system is robustly reliable if it can tolerate a large value of the uncertainty parameter before failure can occur. The basic connection between convex modelling and robust reliability lies in identifying the expansion parameter as a measure of uncertainty, and then evaluating reliability as the amount of acceptable uncertainty.

A convex model $\mathcal{U}(\alpha)$ such as eq.(2.21) is a precisely defined set if α is given a value; any function $u(t)$ either belongs to $\mathcal{U}(\alpha)$ or it does not. However, in robust reliability analysis we will usually deal with the family of sets, $\mathcal{U}(\alpha)$ for $\alpha \geq 0$. Every function (or at least a very broad class of functions) belongs to the family for some value of α. The family of convex models arranges the space of functions in a particular order which depends on the structure of the underlying set whose "shape" generates the family.

In practice we often will not know the value of α, and in robust reliability we do not need to know its value. What we need to know is how the uncertain events cluster, what is the rule which defines their common origin, what is the shape of the convex model, but not how tightly the uncertain events aggregate in any particular instance. Instead, robust reliability analysis focuses on the question: how large a value of α is consistent with no failure.

2.4 The Structure of Convex Sets

We are now ready to describe the quantitative properties of convex sets which will be useful in our study of robust reliability.

2.4.1 Definition of Convexity

A region R in the euclidean plane is convex if the line segment joining any two points in R is entirely in the region. For instance, ovals, triangles and squares delimit convex regions, though quadrilaterals may or may not, as seen in fig. 2.1.

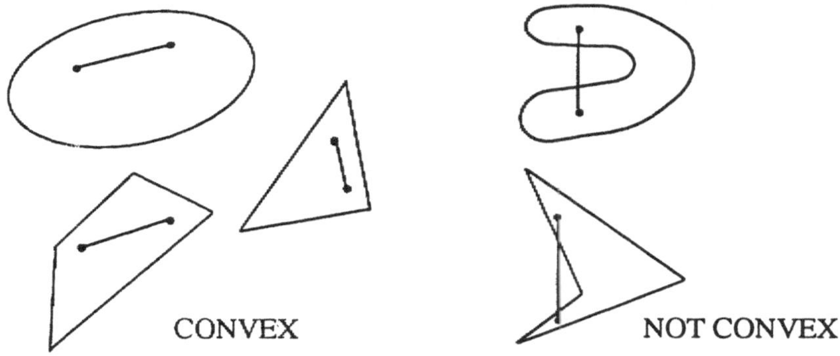

Figure 2.1: Convex and non-convex regions on the plane.

This graphical definition of convexity can be immediately generalized algebraically. Let $\mathcal{S}$ be a set of points in N-dimensional euclidean space $\Re^N$. The set $\mathcal{S}$ is convex if all weighted averages of points in $\mathcal{S}$ also belong to $\mathcal{S}$. That is, $\mathcal{S}$ is convex if, for any points x and y in $\mathcal{S}$, and for any number $0 \leq \beta \leq 1$, the point $\beta x + (1-\beta)y$ also belongs to $\mathcal{S}$:

$$x, y \in \mathcal{S} \quad \text{and} \quad 0 \leq \beta \leq 1 \quad \Longrightarrow \quad \beta x + (1-\beta)y \in \mathcal{S} \tag{2.24}$$

This definition includes the line-segment definition for convex regions on the plane since the line segment joining any to points x and y on the plane is the set of points:

$$\beta x + (1-\beta)y, \quad \text{for all} \quad 0 \leq \beta \leq 1 \tag{2.25}$$

An expression such as $\beta x + (1-\beta)y$, for $0 \leq \beta \leq 1$, is called a *convex combination* of x and y. Similarly, for points $x_1, \ldots, x_m$ and non-negative numbers $\beta_1, \ldots, \beta_m$ which sum to unity, the quantity $\sum_{i=1}^{m} \beta_i x_i$ is the convex combination of these points, and $\beta_1, \ldots, \beta_m$ are *convex coefficients.*

The algebraic definition of convexity, eq.(2.24), is perfectly valid also when the "points" x and y are functions rather than points in euclidean space.

Example 1 Consider the set of scalar functions:

$$\mathcal{S} = \left\{ u(t) : \int_0^\infty u^2(t)\,\mathrm{d}t \leq 1 \right\} \tag{2.26}$$

We can readily show that, for any functions $u(t) \in \mathcal{S}$ and $v(t) \in \mathcal{S}$, and for any $0 \leq \beta \leq 1$, the function $w(t) = \beta u(t) + (1-\beta)v(t)$ also belongs to $\mathcal{S}$. In other words, $\mathcal{S}$ is a convex set. To do this we must show that:

$$\int_0^\infty w^2(t)\,\mathrm{d}t \leq 1 \tag{2.27}$$

which establishes the membership of $w(t)$ in the set $\mathcal{S}$. The Schwarz inequality states that, for any real functions $x(t)$ and $y(t)$, the following inequality holds:

$$\left(\int x(t)y(t)\,\mathrm{d}t\right)^2 \leq \int x^2(t)\,\mathrm{d}t \int y^2(t)\,\mathrm{d}t \tag{2.28}$$

Equality holds in this relation if and only if the functions $x(t)$ and $y(t)$ are proportional to one another:

$$x(t) = cy(t) \tag{2.29}$$

where c is a constant.

The integral in eq.(2.27) can be expressed:

$$\int_0^\infty w^2(t)\,\mathrm{d}t = \int_0^\infty [\beta x(t) + (1-\beta)y(t)]^2\,\mathrm{d}t \tag{2.30}$$

$$= \beta^2 \int_0^\infty x^2(t)\,\mathrm{d}t + 2\beta(1-\beta)\int_0^\infty x(t)y(t)\,\mathrm{d}t + (1-\beta)^2\int_0^\infty y^2(t)\,\mathrm{d}t \tag{2.31}$$

Applying the Schwarz inequality to the second integral on the right this becomes:

$$\int_0^\infty w^2(t)\,\mathrm{d}t \leq \beta^2\int_0^\infty x^2(t)\,\mathrm{d}t + 2\beta(1-\beta)\sqrt{\int_0^\infty x^2(t)\,\mathrm{d}t}\sqrt{\int_0^\infty y^2(t)\,\mathrm{d}t} + (1-\beta)^2\int_0^\infty y^2(t)\,\mathrm{d}t \tag{2.32}$$

Since x and y both belong to $\mathcal{S}$, each integral on the right is no greater than unity. Thus relation (2.32) becomes:

$$\int_0^\infty w^2(t)\,\mathrm{d}t \leq \beta^2 + 2\beta(1-\beta) + (1-\beta)^2 = [\beta + (1-\beta)]^2 = 1 \tag{2.33}$$

which proves that $w(t)$ belongs to $\mathcal{S}$, which is therefore a convex set of functions. ∎

Example 2 We now consider a similar example, this time for N-vectors in euclidean space. Consider the ellipsoid-bound convex model:

$$\mathcal{V} = \left\{v : v^T W v \leq 1\right\} \tag{2.34}$$

where W is a real, symmetric, positive definite matrix. To show that $\mathcal{V}$ is a convex set we must show that, for any vectors u and v belonging to $\mathcal{V}$, and for any number $0 \leq \beta \leq 1$, the vector $w = \beta u + (1-\beta)v$ also belongs to $\mathcal{V}$. In other words, we must show that:

$$w^T W w \leq 1 \tag{2.35}$$

The Cauchy inequality for vectors states:

$$\left(x^T y\right)^2 \leq \left(x^T x\right)\left(y^T y\right) \tag{2.36}$$

with equality if and only if the vectors x and y are proportional to one another:

$$x = cy \tag{2.37}$$

where c is a scalar constant. Now the quadratic term in (2.35) can be written:

$$\begin{aligned} w^T W w &= [\beta u + (1-\beta)v]^T W [\beta u + (1-\beta)v] &(2.38)\\ &= \beta^2 u^T W u + 2\beta(1-\beta) u^T W v + (1-\beta)^2 v^T W v &(2.39) \end{aligned}$$

The middle term on the right can be written:

$$u^T W v = \left(W^{1/2} u\right)^T \left(W^{1/2} v\right) \tag{2.40}$$

Combining this with eq.(2.39) and applying the Cauchy inequality one finds:

$$\begin{aligned} w^T W w \leq\; & \beta^2 u^T W u + 2\beta(1-\beta)\sqrt{u^T W u}\sqrt{v^T W v} \\ & +(1-\beta)^2 v^T W v \end{aligned} \tag{2.41}$$

Since u and v both belong to W, each quadratic term in this expression is no greater than unity. Consequently:

$$w^T W w \leq \beta^2 + 2\beta(1-\beta) + (1-\beta)^2 = [\beta + (1-\beta)]^2 = 1 \tag{2.42}$$

We conclude that w belongs to the set $\mathcal{V}$, which is therefore convex. ■

2.4.2 Extreme Points and Convex Hulls

A square region on the plane is a convex set of points. For example, the following set defines a square centered at the origin:

$$\mathcal{S} = \{(x, y) : \; |x| \leq 1, \;\; |y| \leq 1\} \tag{2.43}$$

Any two points in the region define a line segment lying entirely within the region, hence $\mathcal{S}$ is convex. Conversely, any point P in the interior of the square lies between two boundary points A and B, as in fig. 2.2. Algebraically speaking, any point P in the interior can be expressed as the weighted average

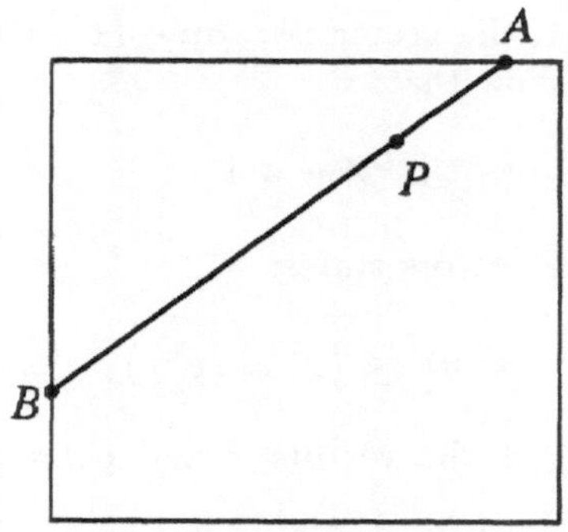

Figure 2.2: The interior point P is the weighted average of boundary points A and B.

of two boundary points A and B, since the coordinates P_x and P_y of P can be expressed as averages of the coordinates of A and B:

$$P_x = \beta A_x + (1-\beta)B_x, \quad P_y = \beta A_y + (1-\beta)B_y \tag{2.44}$$

The same convex coefficients, β and $1-\beta$, specify P_x and P_y. Furthermore, any point along an edge, other than a vertex, is the weighted average of two vertices. Only the vertices themselves cannot be represented as convex combinations of other elements of the set. For any closed, convex set, its *extreme points* are those elements which cannot be expressed as convex combinations of other elements of the set. The extreme points of $\mathcal{S}$ in eq.(2.43) are the four vertices:

$$\mathcal{E} = \{(1,1),\ (1,-1),\ (-1,1),\ (-1,-1)\} \tag{2.45}$$

The *convex hull* of a set $\mathcal{G}$ is the intersection of all convex sets containing $\mathcal{G}$ and is denoted $\mathrm{ch}(\mathcal{G})$. Roughly speaking, $\mathrm{ch}(\mathcal{G})$ is the "smallest" convex set containing $\mathcal{G}$. Every set, whether or not it is convex, has a convex hull. The convex hull of a convex set is the set itself.

Example 3 Consider an ellipsoidal shell in $\Re^N$:

$$\mathcal{E} = \left\{ x : \sum_{n=1}^{N} \frac{x_n^2}{\rho_n^2} = 1 \right\} \tag{2.46}$$

where $\rho_1, \ldots, \rho_N$ are constants. This set is not convex, since the interior of the shell does not belong to the set. However, the solid ellipsoid:

$$\mathcal{S} = \left\{ x : \sum_{n=1}^{N} \frac{x_n^2}{\rho_n^2} \le 1 \right\} \tag{2.47}$$

is convex, contains $\mathcal{E}$, and no proper subset of $\mathcal{S}$ contains $\mathcal{E}$, so $\mathcal{S} = \mathrm{ch}(\mathcal{E})$. Furthermore, $\mathcal{E}$ is the set of extreme points of $\mathcal{S}$. ∎

2.4.3 Extrema of Linear Functions on Convex Sets

Matrix multiplication has the effect of transforming one vector into another:

$$x \longrightarrow y: \quad y = Ax \tag{2.48}$$

Likewise, the integral convolution transforms one function into another:

$$g(t) \rightarrow h(t): \quad h(t) = \int_0^t K(t-\tau)g(\tau)\,\mathrm{d}\tau \tag{2.49}$$

These are examples of linear transformations. A transformation T is *linear* on a set $\mathcal{S}$ if, for all elements x and y in $\mathcal{S}$ and all numbers β and γ, the following equality holds:

$$T(\beta x + \gamma y) = \beta T(x) + \gamma T(y) \tag{2.50}$$

A linear transformation is sometimes called a linear function, linear operator or linear map.

In many analyses of robust reliability we will seek the extrema of a linear function on a convex set. The following theorem will often simplify the analysis.

Theorem 1 *If T is a linear function and $\mathcal{S}$ is a closed and bounded set,*[2] *then T assumes the same extrema on $\mathcal{S}$ and on* ch($\mathcal{S}$).

If $\mathcal{S}$ is a convex set and $\mathcal{E}$ is its set of extreme points, then $\mathcal{S} = \mathrm{ch}(\mathcal{E})$. Consequently, the extrema of a linear function on $\mathcal{S}$ are the same as the extrema of the linear function on $\mathcal{E}$.

Example 4 Let $\mathcal{S}$ be the square region defined in eq.(2.43) and let $\mathcal{E}$ be the set of extreme points, eq.(2.45). Consider the linear function:

$$f(x, y) = x + y \tag{2.51}$$

The maximum of f on $\mathcal{S}$ equals the maximum on $\mathcal{E}$, which is easily found since $\mathcal{E}$ contains only four elements. Thus:

$$\max_{(x,y)\in\mathcal{S}} f(x, y) = \max_{(x,y)\in\mathcal{E}} f(x, y) = f(1, 1) = 2 \tag{2.52}$$

■

Example 5 Consider the ellipsoidal shell defined in eq.(2.46), which we re-write as:

$$\mathcal{E} = \left\{x : \; x^T R x = 1\right\} \tag{2.53}$$

---

[2]This theorem is true for the wider class of compact sets.

where $x \in \Re^N$ and R is the $N \times N$ diagonal matrix with $1/\rho_n^2$ in the nth diagonal position. The solid ellipsoid of eq.(2.47) is the set:

$$\mathcal{S} = \{x : x^T R x \leq 1\} \tag{2.54}$$

We seek the extrema on $\mathcal{S}$ of the following linear function:

$$f(x) = w^T x \tag{2.55}$$

where w is a constant vector. By theorem 1, it is sufficient to seek the extrema of f on $\mathcal{E}$, since $\mathcal{S}$ is the convex hull of $\mathcal{E}$. This is convenient since $\mathcal{S}$ is defined with an inequality, while $\mathcal{E}$ has an equality, so it is easier to optimize f on $\mathcal{E}$ than on $\mathcal{S}$.

We use the method of Lagrange multipliers. We wish to optimize $f(x)$ subject to the constraint:

$$0 = 1 - x^T R x \tag{2.56}$$

Adjoining the constraint to $f(x)$ with an unknown multiplier, λ, we define a new function:

$$J = w^T x + \lambda(1 - x^T R x) \tag{2.57}$$

Optimizing J while also satisfying the constraint, eq.(2.56), is the same as optimizing $f(x)$ subject to the constraint, since the constraint implies that the quantity multiplying λ in (2.57) is zero. A necessary condition for an extremum is:

$$0 = \frac{\partial J}{\partial x} \tag{2.58}$$

$$= w - 2\lambda R x \tag{2.59}$$

Hence the optimizing vector is:

$$x = \frac{1}{2\lambda} R^{-1} w \tag{2.60}$$

This vector must satisfy the constraint, eq.(2.56), so λ must assume one of the following values:

$$\lambda = \pm\frac{1}{2}\sqrt{w^T R^{-1} w} \tag{2.61}$$

Combining the last two relations results in the extrema of $f(x)$ on $\mathcal{S}$:

$$\operatorname*{opt}_{x \in \mathcal{S}} f(x) = \mp\sqrt{w^T R^{-1} w} \tag{2.62}$$

■

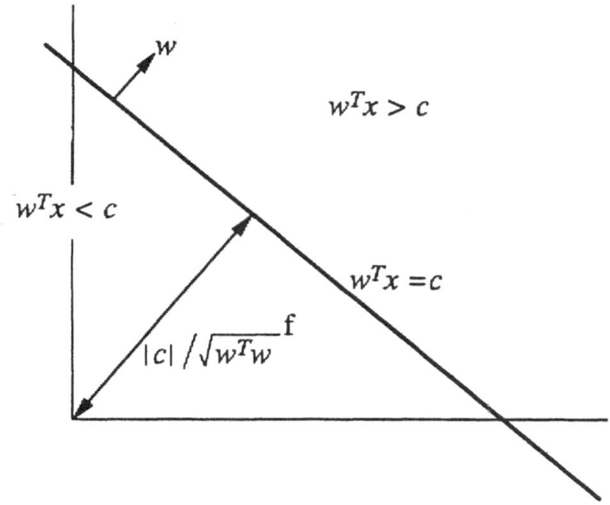

Figure 2.3: Hyperplane and half-spaces defined by a linear function.

2.4.4 Hyperplane Separation of Convex Sets

The analysis of robust reliability will lead to the need to determine the disjointness or intersection of convex sets. This is readily accomplished based on the idea of hyperplane separation, which we now discuss.

Let w be a constant N-vector, so that $f(x) = w^T x$ is a linear function which maps points from $\Re^N$ to $\Re^1$. For any constant, c, a *hyperplane* in $\Re^N$ is the set of points x satisfying:

$$w^T x = c \tag{2.63}$$

For example, if $N = 2$, this is the equation of a straight line in the plane.

The perpendicular distance of the hyperplane (2.63) from the origin is $|c|/\sqrt{w^T w}$. Consequently, if $|c_1| < |c|$, then the hyperplane $w^T x = c_1$ is parallel to the hyperplane $w^T x = c$ and closer to the origin. One implication of this is that all the points on one side of the hyperplane $w^T x = c$ satisfy:

$$w^T x < c \tag{2.64}$$

while all the points on the other side satisfy:

$$w^T x > c \tag{2.65}$$

Thus the hyperplane $w^T x = c$ divides the space into two half-spaces whose members are characterized by the sign of $c - w^T x$, as shown in fig. 2.3.

A closed half-space is the set of points on one side of a hyperplane, including the hyperplane itself. The "stone grinder's" theorem establishes a close relation between convex sets and closed half-spaces.

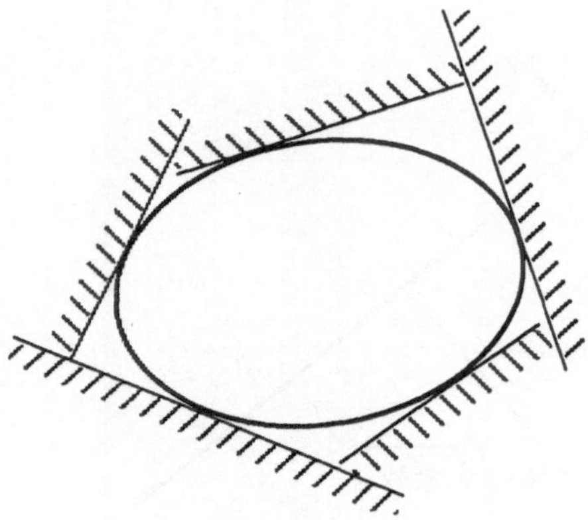

Figure 2.4: Illustration of theorem 2, the stone grinder's theorem.

Theorem 2 *[78, p.99]. A closed convex set is the intersection of all the closed half-spaces which contain it.*

The theorem is illustrated in fig. 2.4. This theorem gets its name because it assures us that any convex shape can be cut from a stone on a flat grinding wheel, (assuming the grinder has sufficient patience).

This theorem is important not only for lapidary art. An immediate implication is that closed convex sets are disjoint if and only if a hyperplane separates them [52, p.145]:

Theorem 3 *Let $\mathcal{G}$ and $\mathcal{H}$ be non-empty, closed convex sets in $\Re^N$ and let one of them be bounded. $\mathcal{G}$ and $\mathcal{H}$ are disjoint if and only if there is a hyperplane such that $\mathcal{G}$ is in one half-space and $\mathcal{H}$ is in the other.*

This theorem can be expressed algebraically as follows, illustrated by fig. 2.5. $\mathcal{G}$ and $\mathcal{H}$ are disjoint if and only if there is a linear function $f(x) = w^T x$, defining a hyperplane, such that the *maximum* of $f(x)$ on one of the sets is less than the *minimum* of $f(x)$ on the other set:

$$\mathcal{G} \cap \mathcal{H} = \emptyset \tag{2.66}$$

if and only if there is a vector w such that:

$$\max_{g \in \mathcal{G}} w^T g \; < \; \min_{h \in \mathcal{H}} w^T h \tag{2.67}$$

A further simplification is obtained by exploiting theorem 1. Suppose that $\mathcal{G}$ and $\mathcal{H}$ are the convex hulls of sets Γ and Φ, respectively. Then theorem 1 implies that the extrema in (2.67) can be sought on Γ and Φ. Thus a necessary and sufficient condition for disjointness of $\mathcal{G}$ and $\mathcal{H}$ is the existence of a vector v such that:

$$\max_{\gamma \in \Gamma} v^T \gamma \; < \; \min_{\phi \in \Phi} v^T \phi \tag{2.68}$$

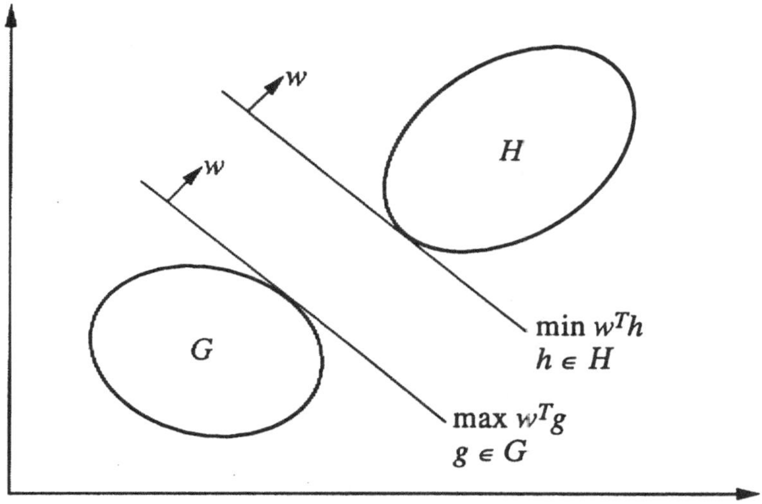

Figure 2.5: Illustration of theorem 3 and eq.(2.67).

2.4.5 Linear Systems Driven by Convex Input Sets

Vibrating mechanical systems such as bridges, buildings, milling machines, surface or air vehicles and so on, are excited by forces which are usually imperfectly known. It is frequently convenient to represent these uncertain loads with convex models. The dynamic vibrations of the mechanical system can often be represented by linear mathematical models, such as state space models or modal-decomposition models. An extremely important property of such representations is that the convex input set generates a convex output set.

Consider the state-space representation of a linear system, where $x(t) \in \Re^N$ is the state vector and $u(t) \in \Re^{N_i}$ is the input or load vector:

$$\frac{\mathrm{d}x}{\mathrm{d}t} = Ax(t) + Bu(t) \tag{2.69}$$

where A and B are constant matrices. The response at time t to load history $u(t)$ is:

$$x_u(t) = \mathrm{e}^{At}x(0) + \int_0^t \mathrm{e}^{A(t-\tau)}Bu(\tau)\,\mathrm{d}\tau \tag{2.70}$$

Let $\mathcal{U}(\alpha)$ represent a convex model of uncertain input functions. That is,

$$u(t) \in \mathcal{U}(\alpha) \tag{2.71}$$

and $\mathcal{U}(\alpha)$ is a convex set of functions. Each input function $u(t)$ generates a response, $x_u(t)$. The *response set* is the set of all the functions $x_u(t)$:

$$\mathcal{R}(\alpha) = \{x(t) : \ x(t) = x_u(t), \quad \text{for all} \quad u(t) \in \mathcal{U}(\alpha)\} \tag{2.72}$$

The integral convolution in eq.(2.70) is a linear operator, so it is readily shown that $\mathcal{R}(\alpha)$ is a convex set since $\mathcal{U}(\alpha)$ is convex (problem 11).

2.5 Clustering of Uncertain Events: The Convexity Theorem

The convex models of uncertainty surveyed in section 2.2 are all *convex sets* of functions. From the engineering point of view, each of these sets is defined as the collection of all elements consistent with given information (an energy bound, a spectral envelope, and so on). The convexity of these sets arises 'naturally', as a by-product. In this section we will discuss a theorem which indicates a connection between convexity and uncertainty.

It is clear that set-models of uncertainty are not by necessity convex; important situations arise in which the sets involved are not convex. However, the following elementary theorem [20] provides some indication of why convexity is not just accidental in the modelling of uncertainty.

Let $f(t)$ be a time (or space) varying uncertain vector function, and let Γ be a set of such functions. For a positive integer n, consider the set of functions constructed as n-fold averages of elements of Γ:

$$\mathcal{F}_n = \left\{ f : \ f(t) = \frac{1}{n} \sum_{m=1}^{n} g_m(t), \quad \text{for all } g_m \in \Gamma \right\} \tag{2.73}$$

It is well known that, as $n \to \infty$, the sequence of sets $\mathcal{F}_n$, $n = 1, 2, \ldots$ converges to the convex hull of Γ:

$$\lim_{n \to \infty} \mathcal{F}_n = \text{ch}\,(\Gamma) \tag{2.74}$$

(For more general results see [3, 4, 5]).

This theorem suggests the following physical interpretation. If an uncertain, macroscopic process, represented by $f(t)$, is formed as the linear superposition of numerous microscopic processes $g_m(t)$, each drawn from the set Γ, then the set of all processes $f(t)$ will tend to be convex, regardless of the structure of the set Γ. In other words, we might expect that complex vector processes will tend to cluster in convex sets of functions. This highlights the distinctive feature of convex models of uncertainty, in contrast to probabilistic models. While probabilistic models emphasize the *frequency of occurrence* of events, convex models stress the *clustering* or *aggregation* of uncertain events.

Eq.(2.74) bears a suggestive similarity to the central limit theorem, even though the contents and proofs of these two theorems differ utterly. Let $g_1, g_2, \ldots$ be independent, identically distributed random variables. Technicalities aside, the central limit theorem states that, as $n \to \infty$, the distribution of the sum $f = \frac{1}{\sqrt{n}} \sum_{m=1}^{n} g_m$ converges to a normal distribution, regardless of how the g_m's are distributed. The physical implication is that if a macroscopic random quantity f is composed of a multitude of superimposed independent random microscopic quantities g_m, then f should tend to display a normal distribution.

The central limit theorem and eq.(2.74) both relate fairly arbitrary microscopic uncertainties to rather more specific macroscopic uncertainty models. Despite this similarity, however, the points of emphasis of these two results are completely disparate. The central limit theorem directs attention to the structure of the *probability measure,* while eq.(2.74) focusses on the structure of the *event set* [16].

Historically speaking, the proof of the central limit theorem, presented in 1810 by Laplace, was a great advance in understanding the fundamental mathematical nature of probability densities [91]. In addition, the theorem provided a justification of the least-squares estimation method developed five years before by Legendre and, independently, by Gauss. Moreover, the central limit theorem directed the attention of researchers to probability densities and their derivations. This led, in the course of the 19th century, to the discovery of other statistical distributions.

In recent decades attention has been placed on extending the concept of probability density. In the place of classical probability functions one has membership functions and measures of possibility and necessity in fuzzy logic [35], belief functions in the Dempster-Shafer theory [75], and so on. The logical diversity of these theories is real and substantial, as evidenced by the distinct axiomatic bases on which they rest [43]. However, the intellectual connection to traditional uncertainty models is clear: modern as well as classical thought concentrates on the properties and structure of normalized non-negative functions defined on sets of events.

In contrast, set-theoretical models of uncertainty, such as convex models, concentrate on the geometric structure of event-clusters. What has attracted the attention of workers in various technological areas is the fact that fragmentary information about uncertain events often leads to the definition of a *convex* set of events. This provides both a standardized framework for analysis, as well as a guide to the formulation of plausible uncertainty models based upon severe lack of prior information.

2.6 Problems

1. Show that any hyperplane in $\Re^N$ is a convex set.

2. Show that eqs.(2.2), (2.5), (2.8) and (2.12) define convex sets.

3. Let C be a convex set and let β be a number. The set obtained by multiplying each element of C by β is denoted βC. Show that βC is convex.

4. Let $\mathcal{S}_1$ be the square centered at the origin, eq.(2.43), and let $\mathcal{S}_2$ be the shifted square:

$$\mathcal{S}_2 = \left\{ (x, y) : \left| x - \frac{1}{2} \right| \leq 1, \ \left| y - \frac{1}{2} \right| \leq 1 \right\} \tag{2.75}$$

Define $\mathcal{S}_3$ as the intersection of $\mathcal{S}_1$ and $\mathcal{S}_2$: $\mathcal{S}_3 = \mathcal{S}_1 \cap \mathcal{S}_2$. Is $\mathcal{S}_3$ convex? Why? What are the extreme points of $\mathcal{S}_3$?

5. Consider the set of N-dimensional functions defined in eq.(2.21). What are the extreme points of this set?

6. What is the qualitative difference between the elements of the cumulative and instantaneous energy-bound convex models, eqs.(2.3) and (2.4)?

7. ‡ Optimize the quadratic function $v^T v$ on the ellipsoid-bound convex model, eq.(2.5). (Hint: this leads to an eigenvalue equation.)

8. Let $\psi(t)$ be a known N-vector function, and optimize the linear integral operator:

$$f(t) = \int_0^T \psi^T(t - \tau) u(\tau) \, \mathrm{d}\tau \tag{2.76}$$

where the N-vector function $u(t)$ belongs to the instantaneous energy-bound convex model, eq.(2.4). (Hint: use the Cauchy inequality).

9. Repeat problem 8 when $u(t)$ belongs to the cumulative energy-bound convex model, eq.(2.3).

10. Consider the following two sets of N-vectors:

$$\mathcal{S}_1(\alpha) = \left\{ x : x^T x \leq \alpha \right\}, \tag{2.77}$$

$$\mathcal{S}_2(\alpha) = \left\{ x : (x - \overline{x})^T (x - \overline{x}) \leq \alpha \right\} \tag{2.78}$$

Determine the smallest value of α such that these sets intersect.

11. Show that the response set $\mathcal{R}(\alpha)$ defined in eq.(2.72) is convex.

12. A diagnostic measurement point on a milling machine vibrates approximately as a simple harmonic oscillator, for which the maximum acceleration, a, is related to the maximum displacement, x, and the square of the natural frequency, ω^2, as $a = \omega^2 x$. Thus, measuring a and x allows one to determine the natural frequency, which is useful diagnostic information. However, a range of complications such as measurement noise, system non-linearities, effects of additional modes of vibration, extraneous excitations and so on cause the measurement (x, a) to vary on a set of values for a given central frequency, ω. In other words,

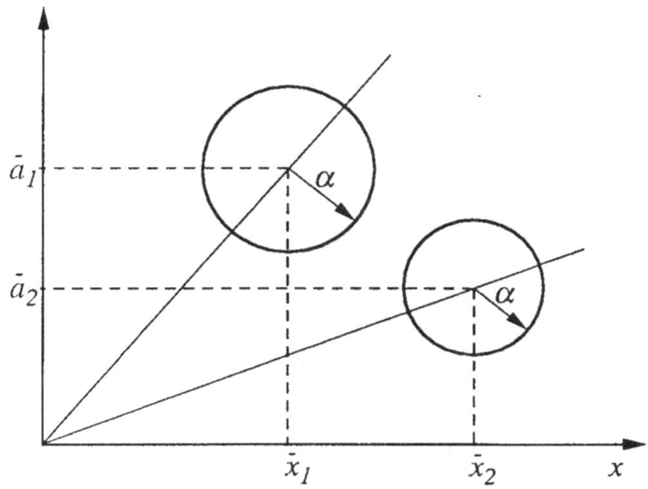

Figure 2.6: Figure for problem 12.

measurement of x and a determines ω^2 with some uncertainty. A simple convex model represents the uncertainty in (x, a) by a circle, as in fig. 2.6. The set of measurements obtainable for given ω^2 is:

$$\mathcal{U}(\alpha, \omega^2) = \left\{(x, a) : \ (x - \overline{x})^2 + (a - \overline{a})^2 \leq \alpha^2\right\} \tag{2.79}$$

where $\overline{a} = \omega^2 \overline{x}$. The parameter α expresses the amount of uncertainty. (a) The measurement (x, a) will be used to distinguish between two conditions of the machine, which have distinct natural frequencies, ω_1 and ω_2. If the measurements must always correctly distinguish between these two conditions, how much uncertainty, α, can be tolerated? (b) For fixed uncertainty, α, and for a given natural frequency ω_1, what other frequencies ω can always be distinguished based on the measurements (x, a)?

Chapter 3

Robust Reliability of Static Systems

3.1 Introduction

Mechanical systems are hardly ever designed to fail. Failure occurs because the system differs from its nominal design, or because the operational environment changes, or the system is altered in some way, or unanticipated or extraordinary loads are applied. The *robust reliability* of a system is a measure of its resistence to these uncertainties. The system is reliable if it can tolerate a large amount of uncertainty without failing. On the other hand, a system is not reliable if it is fragile with respect to uncertainty; it is unreliable if failure becomes a possibility as a result of small deviations from nominal circumstances.

The analysis of robust reliability rests on three components:

1. A *mechanical model*, describing the physical properties of the system.

2. A *failure criterion*, specifying the conditions which constitute failure of the system.

3. An *uncertainty model*, quantifying the uncertainties to which the system is subjected. These uncertainties may relate to the operational environment of the system, and may appear in the mechanical model as well as in the failure criterion.

By combining these three components we determine the robust reliability: the greatest uncertainty which the mechanical system can tolerate without failure.

In this chapter we will analyze the reliability of a range of mechanical systems. We limit ourselves to systems in which the uncertain phenomenon

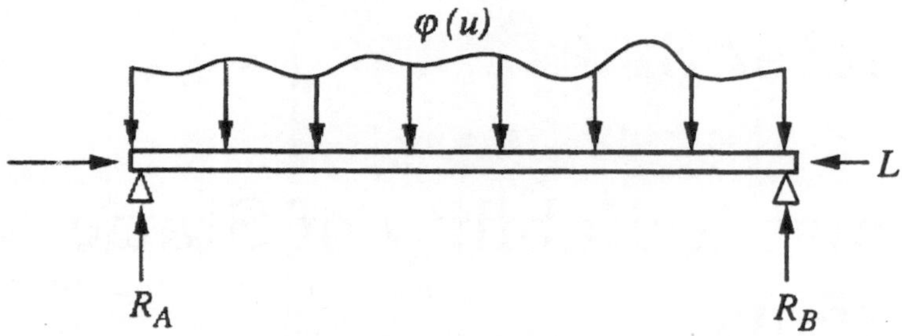

Figure 3.1: Uncertain force profile on a beam.

is not time varying. The systems considered in this chapter are also, by and large, static in time. The basic goal in each case is to determine the robust reliability. However, other questions arise as well.

The analysis of reliability allows one to identify the influence of various properties or parameters of the mechanical system on its reliability. This leads naturally to optimization of the system design with respect to reliability.

The analysis of reliability establishes a connection between uncertainty and reliability. When the uncertain factors, for instance geometrical dimensions or material properties, can be controlled during manufacture, the analysis motivates the definition of criteria for quality control.

We will study examples of these ideas in the sections which follow.

3.2 Beam With An Uncertain Distributed Load

We begin with a simple example to illustrate the basic features of robust reliability analysis. We will repeat this example four times in the present section, each time with a different uncertainty model. The outline of the analysis is the same in each case, but the details of the results can be quite different, depending on the prior information about the uncertainty as reflected in the convex model of the load variation.

3.2.1 Uniform Load Uncertainty

Consider a simply supported uniform beam of length L with a distributed load $\phi(u)$ [N/m], as in fig. 3.1. We must specify the mechanical model, the uncertainty model and the failure criterion. We then determine the robust reliability: the greatest uncertainty the mechanical system can tolerate without failure.

Let R_A denote the reaction at the $x = 0$ end and $M(x)$ the bending moment at position x along the beam. Using the fact that $M(L) = 0$, the bending moment at point x is:

$$M(x) = -R_A x + \int_0^x \phi(u)(x-u)\,\mathrm{d}u \tag{3.1}$$

$$= -\frac{x}{L}\int_0^L \phi(u)(L-u)\,\mathrm{d}u + \int_0^x \phi(u)(x-u)\,\mathrm{d}u \tag{3.2}$$

Now let us suppose that the load profile is uncertain, but bounded uniformly along the beam. We will suppose that the cnly information we have about the load is that its magnitude nowhere exceeds some constant unknown value α. We represent the load uncertainty with a uniform-bound convex model:

$$\mathcal{U}(\alpha) = \{\phi(u) : \ |\phi(u)| \le \alpha, \quad 0 \le u \le L\} \tag{3.3}$$

The set $\mathcal{U}(\alpha)$ represents the range of uncertain variation of the load profile, and α, the uncertainty parameter, is a measure of the degree of uncertainty. If α is very small, then the load is nearly zero everywhere along the beam. Or, if α is large then the load *may* be large anywhere or everywhere. The greater the value of α, the less we know about the load profile.

Let our failure criterion be proportional to the bending moment. That is, the beam fails if:

$$\max_x |M(x)| > M_{\mathrm{cr}} \tag{3.4}$$

where M_{cr} is the critical bending moment.

For fixed position x along the beam, we can readily find the greatest bending moment from among all the profiles consistent with the convex model, eq.(3.3). We re-write eq.(3.2) as:

$$M(x) = -\frac{L-x}{L}\int_0^x \phi(u)u\,\mathrm{d}u - \frac{x}{L}\int_x^L \phi(u)(L-u)\,\mathrm{d}u \tag{3.5}$$

The integrands of both integrals have the same sign everywhere, while both integrals have negative coefficients. So, the maximum of $M(x)$, subject to $\phi(u) \in \mathcal{U}(\alpha)$, is obtained by choosing $\phi(u) = -\alpha$ at each point on the beam. Thus:

$$\max_{|\phi(u)|\le\alpha} |M(x)| = -\frac{L-x}{L}\int_0^x (-\alpha)u\,\mathrm{d}u - \frac{x}{L}\int_x^L (-\alpha)(L-u)\,\mathrm{d}u \tag{3.6}$$

$$= \frac{\alpha}{2}x(L-x) \tag{3.7}$$

This is the greatest bending moment at position x, from among all the allowed load profiles bounded by amplitude α. Becausè the set $\mathcal{U}(\alpha)$ of uncertain load profiles is symmetric with respect to zero, the least bending moment is the negative of eq.(3.7).

The maximum along the beam of $x(L-x)$ occurs at $x = L/2$. So, the greatest bending moment which can occur in the beam, subject to the bounded uncertainty in the load profile, is:

$$\max_{0\le x\le L} \max_{|\phi(u)|\le\alpha} |M(x)| = \frac{\alpha}{2}\cdot\frac{L}{2}\cdot\frac{L}{2} = \frac{\alpha L^2}{8} \tag{3.8}$$

The greatest value of the load-uncertainty parameter which is consistent with no-failure will be denoted throughout the book by $\widehat{\alpha}$. In the present example, this critical value of the uncertainty is obtained by comparing relations (3.4) and (3.8):

$$M_{\rm cr} = \frac{\alpha L^2}{8} \quad\Longrightarrow\quad \widehat{\alpha} = \frac{8M_{\rm cr}}{L^2} \tag{3.9}$$

Now we will apply this result to the concept of robust reliability: a structure is reliable if it is robust with respect to uncertainty; it is unreliable if small uncertainty can lead to failure. If $\widehat{\alpha}$ is large then the beam will not fail even in the presence of large load uncertainty. On the other hand, if $\widehat{\alpha}$ is small then failure can occur even with slight deviation of the load profile from the nominal zero value. $\widehat{\alpha}$ is a measure of the reliability of the beam. One implication is that the reliability in this example decreases inversely with the square of the beam length.

3.2.2 Shifted Uniform Load Uncertainty

Now let us suppose that there is nominally a constant positive load $\overline{\phi}$ [N/m] along the beam, subject to uncertainty α. Instead of eq.(3.3) we employ the shifted uniform-bound convex model of load profiles:

$$\mathcal{U}_{\rm s}(\alpha) = \left\{\phi(u) : \ |\phi(u)-\overline{\phi}| \le \alpha, \quad 0\le u\le L\right\} \tag{3.10}$$

In other words, our prior information concerning the load uncertainty is different from the example of section 3.2.1.

We must seek the maximum of $|M(x)|$. However, unlike section 3.2.1, the minimum and maximum moments are not equal in magnitude because the convex model of load profiles is not symmetric with respect to zero. Eq.(3.5) is still valid, and arguing as in eq.(3.6) we find:

$$\max_{\phi(u)\in\mathcal{U}_{\rm s}(\alpha)} M(x) = -\frac{(\alpha-\overline{\phi})x(L-x)}{2} \tag{3.11}$$

$$\min_{\phi(u)\in\mathcal{U}_{\rm s}(\alpha)} M(x) = -\frac{(\alpha+\overline{\phi})x(L-x)}{2} \tag{3.12}$$

So, instead of eq.(3.7) we have:

$$\max_{\phi(u)\in\mathcal{U}_{\rm s}(\alpha)} |M(x)| = \frac{(\alpha+\overline{\phi})x(L-x)}{2} \tag{3.13}$$

and instead of eq.(3.8) we find the greatest bending moment, which occurs at the midpoint of the beam, to be:

$$\max_{0\leq x\leq L} \max_{\phi(u)\in\mathcal{U}_s(\alpha)} |M(x)| = \frac{(\alpha+\overline{\phi})L^2}{8} \tag{3.14}$$

Applying the failure criterion, eq.(3.4), we evaluate the robust reliability by equating the maximum moment to the critical moment and then solving for α:

$$M_{\rm cr} = \frac{(\alpha+\overline{\phi})L^2}{8} \quad\Longrightarrow\quad \widehat{\alpha} = \frac{8M_{\rm cr}}{L^2} - \overline{\phi} \tag{3.15}$$

Comparing this with eq.(3.9), we see that the nomimal load $\overline{\phi}$ reduces the robustness of the beam to load uncertainty. Furthermore, the reliability decreases as $\overline{\phi}$ increases, reaching zero when the nominal load produces the critical bending moment. At this level of nominal load even the smallest load deviation could result in failure.

3.2.3 Load-Uncertainty Envelope

Let us suppose that the nominal load is positive but varies along the beam, so the constant $\overline{\phi}$ of eq.(3.10) now becomes a function, $\overline{\phi}(u)$. Also, let us presume additional prior information which specifies bounds which vary along the beam, implying greater load uncertainty in some parts of the beam than in others. The envelope-bound convex model describes the uncertain load profiles:

$$\mathcal{U}_{\rm e}(\alpha) = \left\{\phi(u):\ |\phi(u)-\overline{\phi}(u)| \leq \alpha\psi(u),\quad 0\leq u\leq L\right\} \tag{3.16}$$

As before, α represents the "size" of the convex model and the overall load uncertainty. When α is small the load profile is close to the nominal profile $\overline{\phi}(u)$, while as α increases the range of variation of $\phi(u)$ increases as well. The functions $\overline{\phi}(u)$ and $\psi(u)$ are known and are derived from our prior information about the load uncertainty. The value of the uncertainty parameter α need not be known. This convex model is more detailed than the uniform-bound convex model $\mathcal{U}_s(\alpha)$ of eq.(3.10).

Eq.(3.5) describes the bending moment at position x on the beam. To find the maximum absolute value of the bending moment we evaluate the minimum and maximum of $M(x)$. Because of the negative coefficients of the integrals in eq.(3.5), the minimum is obtained by choosing the load to be maximal at each point, $\phi(u) = \overline{\phi}(u) + \alpha\psi(u)$, resulting in:

$$\begin{aligned} M_1(x) &= \min_{\phi(u)\in\mathcal{U}_e} M(x) && (3.17)\\ &= -\frac{L-x}{L}\int_0^x [\overline{\phi}(u)+\alpha\psi(u)]u\,{\rm d}u \\ &\quad -\frac{x}{L}\int_x^L [\overline{\phi}(u)+\alpha\psi(u)](L-u)\,{\rm d}u && (3.18) \end{aligned}$$

The maximum bending moment is obtained by using the minimum load profile, $\phi(u) = \overline{\phi}(u) - \alpha\psi(u)$, leading to:

$$M_2(x) = \max_{\phi(u)\in\mathcal{U}_e} M(x) \tag{3.19}$$

$$= -\frac{L-x}{L}\int_0^x [\overline{\phi}(u) - \alpha\psi(u)]u\,du - \frac{x}{L}\int_x^L [\overline{\phi}(u) - \alpha\psi(u)](L-u)\,du \tag{3.20}$$

These extremal moments can be succintly expressed as follows:

$$M_1(x) = \eta_1(x) + \alpha\eta_2(x) \tag{3.21}$$

$$M_2(x) = \eta_1(x) - \alpha\eta_2(x) \tag{3.22}$$

where we have defined:

$$\eta_1(x) = -\frac{L-x}{L}\int_0^x u\overline{\phi}(u)\,du - \frac{x}{L}\int_x^L (L-u)\overline{\phi}(x)\,du \tag{3.23}$$

$$\eta_2(x) = -\frac{L-x}{L}\int_0^x u\psi(u)\,du - \frac{x}{L}\int_x^L (L-u)\psi(x)\,du \tag{3.24}$$

Example 1 We continue with an example based on a specific choice of the positive nominal load profile $\overline{\phi}(u)$ and the envelope function $\psi(u)$. Suppose that the nominal load increases toward the center of the beam as a half-sine wave:

$$\overline{\phi}(u) = \overline{\phi}\sin\frac{\pi u}{L} \tag{3.25}$$

where $\overline{\phi} > 0$. Also, let the uncertainty envelope increase similarly towards the middle of the beam:

$$\psi(u) = \sin\frac{\pi u}{L} \tag{3.26}$$

The load functions $\phi(u)$ and $\overline{\phi}(u)$ and the uncertainty parameter α all have the same units, while the envelope $\psi(u)$ is dimensionless in this example. The functions $\eta_1(x)$ and $\eta_2(x)$ become:

$$\eta_1(x) = -\frac{L^2\overline{\phi}}{\pi^2}\sin\frac{\pi x}{L} \tag{3.27}$$

$$\eta_2(x) = \eta_1(x)/\overline{\phi} \tag{3.28}$$

The least and greatest bending moment at point x on the beam can now be written as:

$$M_1(x) = -(\overline{\phi} + \alpha)\frac{L^2}{\pi^2}\sin\frac{\pi x}{L} \tag{3.29}$$

$$M_2(x) = -(\overline{\phi} - \alpha)\frac{L^2}{\pi^2}\sin\frac{\pi x}{L} \tag{3.30}$$

The greatest absolute bending moment occurs at the midpoint of the beam, and is:

$$\max_x |M(x)| = \frac{(\overline{\phi} + \alpha)L^2}{\pi^2} \tag{3.31}$$

The robust reliability is obtained by equating the maximum bending moment to the critical value and solving for the uncertainty parameter α:

$$M_{\rm cr} = \frac{(\overline{\phi} + \alpha)L^2}{\pi^2} \quad \Longrightarrow \quad \widehat{\alpha} = \frac{\pi^2 M_{\rm cr}}{L^2} - \overline{\phi} \tag{3.32}$$

This result is quite similar to eq.(3.15), the difference arising from the different shapes of the nominal load and the envelope function. ∎

3.2.4 Fourier Ellipsoid-Bound Uncertainty

The convex models (3.3), (3.10) and (3.16) employed in the last three subsections share an important common property, namely, that they impose no constraint on the rate of variation of the load profile along the beam. These uncertainty-sets all contain load profiles which vary quite rapidly along the beam, as well as profiles which change very slowly. For example, these sets all contain step-wise varying functions, as well as functions with little or no deviation from the nominal profile. We may have prior information about the physical processes generating the loads, which would lead us to exclude such load profiles. It not infrequently happens that information is available which indicates which spatial frequencies occur, and what ranges of amplitudes they may assume. For instance, in wind or wave excitation problems, or seismic applications, either experimental data or theoretical considerations can provide information of this sort. The Fourier ellipsoid-bound convex model provides one method for representing load uncertainty with this information, and we will now demonstrate its use in robust reliability analysis.

Let us suppose that the distributed load $\phi(u)$ can be represented with the following truncated Fourier series. That is, only particular spatial frequencies contribute to the load profiles:

$$\phi(u) = \sum_{n=n_1}^{n_2} \beta_n \sin \frac{n\pi u}{L}, \quad 0 \le u \le L \tag{3.33}$$

where β_n is the uncertain Fourier coefficient of the nth mode of the load profile. This sum can be represented as a scalar product if we define the vector $\beta^T = (\beta_{n_1}, \ldots, \beta_{n_2})$ of Fourier coefficients and the vector $\gamma^T(u) = (\sin n_1 \pi u/L, \ldots, \sin n_2 \pi u/L)$ of trigonometric functions. Eq.(3.33) becomes:

$$\phi(u) = \beta^T \gamma(u) \tag{3.34}$$

The uncertainty in the load profile is expressed in terms of uncertainty in the Fourier coefficient vector β. The Fourier ellipsoid-bound model constrains

β within an ellipsoid:

$$\mathcal{U}_{\mathrm{f}}(\alpha) = \{\beta : \ \beta^T W \beta \leq \alpha^2\} \tag{3.35}$$

where W is a positive definite real symmetric matrix. Combining eqs.(3.5) and (3.34), the bending moment at position x on the beam becomes:

$$M(x) = \beta^T \underbrace{\left[-\frac{L-x}{L} \int_0^x u\gamma(u)\, \mathrm{d}u - \frac{x}{L} \int_x^L (L-u)\gamma(u)\, \mathrm{d}u \right]}_{\zeta(x)} \tag{3.36}$$

$$= \beta^T \zeta(x) \tag{3.37}$$

which defines the known vector function $\zeta(x)$ which is independent of the load uncertainty.

We can express the bending moment as:

$$\beta^T \zeta(x) = \left(W^{1/2}\beta\right)^T \left(W^{-1/2}\zeta(x)\right) \tag{3.38}$$

Cauchy's inequality, discussed in chapter 2, allows us to find the maximum bending moment, for all Fourier vectors in $\mathcal{U}_{\mathrm{f}}$:

$$\max_{\beta \in \mathcal{U}_{\mathrm{f}}} \beta^T \zeta(x) = \alpha \sqrt{\zeta^T(x) W^{-1} \zeta(x)} \tag{3.39}$$

We note that the greatest bending moment at position x along the beam is proportional to the uncertainty parameter, α, while the term in the radical is completely known. Applying the bending-moment failure criterion, eq.(3.4), the robust reliability can be expressed as:

$$\widehat{\alpha} = \frac{M_{\mathrm{cr}}}{\max_{0 \leq x \leq L} \sqrt{\zeta^T(x) W^{-1} \zeta(x)}} \tag{3.40}$$

While some numerical analysis is required to find the maximum in the denominator of this expression, this is assisted by the fact that the elements of the vector $\zeta(x)$ can be expressed analytically as:

$$\zeta_n(x) = -\frac{L^2}{n^2\pi^2} \sin \frac{n\pi x}{L} \tag{3.41}$$

Example 2 Suppose W is the identity matrix, so the ellipsoid is simply a sphere. Then:

$$\zeta^T(x)\zeta(x) = \frac{L^4}{\pi^4} \sum_{n=n_1}^{n_2} \frac{1}{n^4} \sin^2 \frac{n\pi x}{L} \tag{3.42}$$

The terms in this sum decrease rapidly with n, so that the maximum on x of $\zeta^T(x)\zeta(x)$ is dominated by the first term:

$$\max_{0 \leq x \leq L} \sqrt{\zeta^T(x)\zeta(x)} \approx \frac{L^2}{n_1^2 \pi^2} \tag{3.43}$$

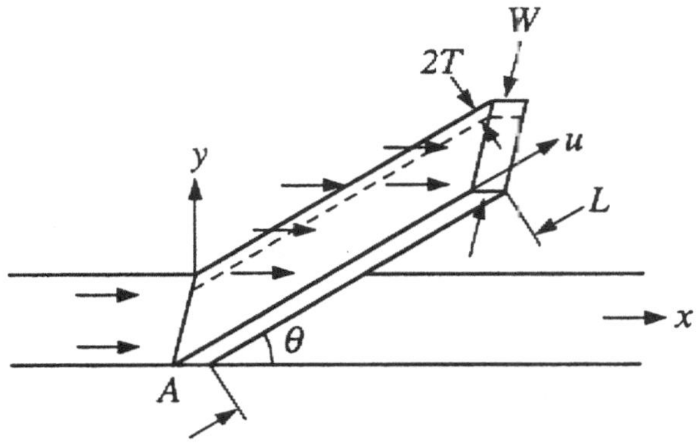

Figure 3.2: Fin configuration.

Referring to eq.(3.40) we see that the reliability is very nearly given by:

$$\widehat{\alpha} \approx \frac{n_1^2 \pi^2 M_{cr}}{L^2} \tag{3.44}$$

Comparing this with the result from the uniform-bound model, eq.(3.9), we see that the reliability is substantially enhanced by constraining the spatial frequencies which compose the load function. ■

3.3 Cooling Fin in an Uncertain Flow Field: Reliability and Design

In this section we will extend our reliability analysis to include consideration of reliable design. Since the expression for robust reliability depends on the design parameters of the structure, we can identify the relative importance of different parameters for reliability. Furthermore, we will discuss the idea of design efficiency for reliability. Rather than having the reliability controlled by a single location on the structure, we choose the design so that the local reliability at each point on the structure is the same.

3.3.1 Uniform Blade

Consider a cooling fin with constant cross section along its length L, with width W and thickness $2T$, and slanted at pitch θ with respect to a flowing fluid impinging on its broad side, as in fig. 3.2. The force imposed by the fluid on the fin is $\phi(y)$ [N/vertical meter]. The fluid force is directed along the x axis, and its amplitude is uncertain and variable along the length of the fin. Our aim is to determine the robust reliability as a function of the geometrical parameters, length and pitch. We must consider the uncertainty model of the fluid force, the mechanical model of the fin, and the criterion by which the fin fails. By combining these three components we determine the robust reliability: the greatest uncertainty which the mechanical system can tolerate without failure.

The uncertainty of the fluid force is represented by a uniform-bound convex model:

$$\mathcal{U}(\alpha) = \{\phi(y) : 0 \le \phi(y) \le \alpha\} \tag{3.45}$$

We will assume that the fin fails if the absolute value of the bending moment at any point v along the fin exceeds a critical value of $M_{\rm cr}$:

$$|M(v)| > M_{\rm cr} \tag{3.46}$$

The mechanical model of the fin expresses the bending moment $M(v)$ at position v along the fin, for fixed force profile $\phi(y)$. To determine $M(v)$ we first determine the reaction force and moment at the root of the fin, point A. The force on an infinitestimal element du along the fin is:

$$dF = \phi(u) \sin\theta \, du \tag{3.47}$$

For force equilibrium we require that the reaction force in the x direction at A be:

$$A_x = \int_0^L dF = \int_0^L \phi(u) \sin\theta \, \mathrm{d}u \tag{3.48}$$

The moment of force at A due to the load is obtained from:

$$M_A = \int_0^L u \sin\theta \, \mathrm{d}F = \int_0^L [u \sin\theta][\phi(u) \sin\theta \, \mathrm{d}u] \tag{3.49}$$

$$= \sin^2\theta \int_0^L u\phi(u) \, \mathrm{d}u \tag{3.50}$$

Now we calculate the bending moment $M(v)$ at a distance v along the fin from A by the method of sections. Equilibrium on the section from A to v requires that the total moment vanish:

$$0 = -A_x v \sin\theta + M_A - M(v) + \int_0^v [(v-u)\sin\theta][\phi(u) \sin\theta \, \mathrm{d}u] \tag{3.51}$$

After some manipulations one finds the moment at v to be:

$$M(v) = \sin^2\theta \int_v^L (v-u)\phi(u) \, \mathrm{d}u \tag{3.52}$$

We must determine the greatest moment, which will depend on the uncertainty parameter, α. Then, by applying the failure criterion we will determine the robust reliability.

The extremal bending moment at v is obtained by choosing $\phi(u)$ in the integral of eq.(3.52) from $\mathcal{U}(\alpha)$ to maximize $|M(v)|$. The term $(v-u)$ in the integrand is negative everywhere, so the maximum of $|M(v)|$ is obtained by choosing $\phi(u)$ constant and equal to α:

$$\widehat{M}(v) = \max_{0\leq\phi(u)\leq\alpha} |M(v)| = \alpha\sin^2\theta \int_v^L (v-u)\,du \tag{3.53}$$

$$= \frac{\alpha}{2}(L-v)^2\sin^2\theta, \quad 0\leq v\leq L \tag{3.54}$$

The maximum bending moment occurs at the root of the fin and is:

$$\widehat{M} = \max_{0\leq v\leq L} \widehat{M}(v) \tag{3.55}$$

$$= \widehat{M}(0) = \frac{\alpha L^2\sin^2\theta}{2} = \frac{\alpha H^2}{2} \tag{3.56}$$

where $H = L\sin\theta$ is the vertical displacement of the fin from root to tip.

We determine the greatest uncertainty which can be tolerated without failure by employing the greatest bending moment in the condition for failure:

$$\widehat{M} = M_{\rm cr} \Longrightarrow \widehat{\alpha} = \frac{2M_{\rm cr}}{L^2\sin^2\theta} \tag{3.57}$$

$$= \frac{2M_{\rm cr}}{H^2} \tag{3.58}$$

From eq.(3.57) we see that, for fixed fin length L, the robustness, $\widehat{\alpha}$, increases as the squared inverse of the sine of the pitch angle: low pitch is very robust, high pitch has low robustness. Eq.(3.58) implies that the robustness decreases with the inverse square of the root-to-tip displacement, H.

$\widehat{\alpha}$ is the greatest load uncertainty the fin can tolerate and not fail anywhere along its length. The failure is however in fact controlled by the bending moment at the root, whose maximum is greater than at any other point on the fin. In other words, positions $v>0$ can tolerate greater uncertainty than $\widehat{\alpha}$ without failing. We could say that $\widehat{\alpha}$ is a conservative reliability estimate. Or, we could say that this fin design is inefficient from the point of view of reliability, since some parts of the fin can tolerate greater uncertainty than other parts.

To explore this further, let us determine the greatest uncertainty tolerable at point v. Failure at v occurs if:

$$\widehat{M}(v) > M_{\rm cr} \tag{3.59}$$

Combining this with eq.(3.54) determines the critical load uncertainty for position v:

$$\widehat{\alpha}(v) = \frac{2M_{\rm cr}}{(L-v)^2\sin^2\theta} \tag{3.60}$$

Comparing this with the overall reliability, eq.(3.57), we see that:

$$\widehat{\alpha}(v) \geq \widehat{\alpha} \tag{3.61}$$

with strict inequality for all $v > 0$.

We see that this rectangular fin cross-section is inefficient from the point of view of reliability: some parts of the fin are much more robust to uncertainty than other parts. In the following subsections we will explore various options for attaining more uniformly reliable design. In section 3.3.2 we will keep the width W constant and determine the optimal fin-thickness profile $T(v)$. In section 3.3.3 we first consider linear variation of the width and find the optimal thickness variation. Alternatively, we find the optimal width profile given parabolic variation of the thickness. Finally, in section 3.3.4 we design both W and T to obtain a minimum-weight uniform-reliability design.

3.3.2 Optimal Thickness Profile

Suppose that the fin has a rectangular cross section with constant width W of the surface on which the fluid impinges, and thickness $2T(v)$ whose variation along the fin length we are free to choose. In the previous section we saw that the bending moment at the root of the fin controlled the reliability, and that the local reliability is not constant along the fin. In this subsection we will investigate the design-for-reliability problem of choosing the thickness profile for maximum efficiency from the point of view of reliability.

The uncertainty model is eq.(3.45), as before. The mechanical model for the bending moment is eq.(3.52), and its maximum on the set of uncertain load profiles is eq.(3.54).

We now define the failure criterion. If the bending moment at v is $M(v)$, then the normal stress (tension or compression) in the cross section, at height t above the neutral plane, is:

$$s(v,t) = \frac{M(v)t}{I(v)} \tag{3.62}$$

where $I = 2WT^3/3$ is the area moment of inertia of the cross section with respect to the neutral plane. So, the greatest normal stress in the fin section at point v along the fin is:

$$s(v) = \frac{M(v)T(v)}{I(v)} = \frac{3M(v)}{2WT^2(v)} \tag{3.63}$$

The fin fails if the normal stress at any point exceeds the critical value of $s_{\rm cr}$:

$$s(v) > s_{\rm cr} \tag{3.64}$$

Our method of analysis is to determine the local robustness $\widehat{\alpha}(v)$: the maximum tolerable uncertainty in the load at position v along the fin. We then will be able to choose the fin profile $T(v)$ for which the robustness $\widehat{\alpha}(v)$ is constant along the fin. This profile is "maximally efficient" from the point of view of reliability.

Using the expression for the maximum bending moment at point v along the fin, eq.(3.54), we find the maximum normal stress on the cross section at point v to be:

$$\widehat{s}(v) = \frac{\widehat{M}(v)T(v)}{I(v)} = \frac{3\alpha(L-v)^2\sin^2\theta}{4WT^2(v)} \tag{3.65}$$

The greatest load uncertainty tolerable at this point on the fin is obtained by requiring that $\widehat{s}(v)$ not exceed the critical stress:

$$\widehat{s}(v) = s_{\text{cr}} \implies \widehat{\alpha}(v) = \frac{4WT^2(v)s_{\text{cr}}}{3(L-v)^2\sin^2\theta} \tag{3.66}$$

From this relation we see that the local value of the critical load uncertainty, $\widehat{\alpha}(v)$, is constant along the fin if the thickness decreases linearly to zero at the tip:

$$T(v) \propto L - v \tag{3.67}$$

So, from the point of view of robustness to load uncertainty, the most efficient fin thickness profile is a triangular shape with zero thickness at the tip.

3.3.3 Optimal Width and Thickness Profiles

Now we increase our design freedom, and allow both the width and thickness of the fin to vary along the length.

Case 1. We first suppose the width to vary linearly and we seek the optimal thickness profile. The given width profile is:

$$W(v) = \beta v + \gamma, \quad 0 \leq v \leq L \tag{3.68}$$

The width has either an increasing or a decreasing taper, depending on the sign of β.

Using the expression for the maximum bending moment at point v along the fin, eq.(3.54), we find the maximum normal stress in the fin cross section at point v, as in eq.(3.65) except that now both W and T depend on position:

$$\widehat{s}(v) = \frac{\widehat{M}(v)T(v)}{I(v)} = \frac{3\alpha(L-v)^2\sin^2\theta}{4W(v)T^2(v)} \tag{3.69}$$

Equating this with the critical stress yields the greatest load uncertainty tolerable at point v, like eq.(3.66):

$$\widehat{s}(v) = s_{\text{cr}} \implies \widehat{\alpha}(v) = \frac{4W(v)T^2(v)s_{\text{cr}}}{3(L-v)^2\sin^2\theta} \tag{3.70}$$

If the width varies linearly along the length of the fin as in eq.(3.68), then the robustness at point v becomes, from eq.(3.70):

$$\widehat{\alpha}(v) = \frac{4(\beta v + \gamma)T^2(v)s_{\rm cr}}{3(L-v)^2 \sin^2\theta} \tag{3.71}$$

For maximum efficiency from the reliability viewpoint, we choose the thickness profile so that $\widehat{\alpha}(v)$ is constant along the length of the fin. So, we require the thickness profile to have the following shape:

$$T(v) \ \propto \ \frac{L-v}{\sqrt{\beta v + \gamma}} \tag{3.72}$$

Note that:

$$\frac{dT}{dv} \propto -\frac{(\beta v + \gamma) + (1/2)\beta(L-v)}{(\beta v+\gamma)^{3/2}} = -\frac{W(v) + (1/2)\beta(L-v)}{W^{3/2}(v)} \tag{3.73}$$

which is not always negative if $\beta < 0$, implying that the most efficient fin-thickness profile is not necessarily monotonically tapered in thickness from root to tip.

Case 2. We now consider the reverse situation: the thickness profile is given and we ask for the most efficient width profile. Let the thickness of the fin vary parabolically along the length as:

$$T^2(v) = \eta(L-v), \quad 0 \le v \le L \tag{3.74}$$

where η is a constant. Find the variation of the width for maximum efficiency or uniform reliability: find $W(v)$ so that $\widehat{\alpha}(v)$ is constant.

Combining eqs.(3.70) and (3.74), the local robustness is:

$$\widehat{\alpha}(v) = \frac{4W(v)\eta s_{\rm cr}}{3(L-v)\sin^2\theta} \tag{3.75}$$

So, a linear width-taper with zero width at the tip is maximally efficient when the thickness is tapered parabolically as in eq.(3.74):

$$W(v) \ \propto \ L - v \tag{3.76}$$

3.3.4 Minimum Weight Design

In the previous subsection we considered the uniform-reliability or maximum-efficiency design of one profile of the fin, either width or thickness, assuming that the other profile was determined beforehand. Let us simultaneously design both the width and the thickness profiles, to obtain a minimum-weight design which is also efficient from the reliability viewpoint.

Assuming uniform density throughout the fin, we will minimize the volume. The fin, whose width is $W(v)$ and thickness is $2T(v)$, has volume:

$$V = 2\int_0^L W(v)T(v)\,\mathrm{d}v \tag{3.77}$$

Referring to eq.(3.70) we recall that the width and thickness profiles for uniform reliability along the length of the fin must be related as follows:

$$W(v)T^2(v) \;\propto\; (L-v)^2 \tag{3.78}$$

A great variety of profiles would satisfy this constraint. From a delimited class of profiles, we will find that particular profile which also minimizes the volume of the fin. Consider only polynomial profiles of the form:

$$W(v) = \beta\left(\frac{L-v}{L}\right)^{\gamma}, \quad T(v) = \eta\left(\frac{L-v}{L}\right)^{\xi} \tag{3.79}$$

The shapes of the profiles are determined by the dimensionless exponents, γ and ξ, which must both be positive to keep W and T finite.

The uniform-reliability requirement, eq.(3.78), implies that the exponents γ and ξ are related as:

$$\gamma + 2\xi = 2 \quad \text{or} \quad \gamma = 2(1-\xi) \tag{3.80}$$

Thus, since $\gamma \geq 0$, we see that $0 \leq \xi \leq 1$. Now eq.(3.77) can be integrated to find the volume as a function of the unknown exponent, ξ:

$$V = \frac{2\beta\eta L}{3-\xi} \tag{3.81}$$

The volume is minimized by choosing $\xi = 0$ and $\gamma = 2$. Thus the uniformly reliable fin of minimum volume, with width and thickness profiles of eq.(3.79), has constant thickness and quadratically varying width.

3.3.5 Parameter Sensitivity of the Reliability

In the previous subsections we have derived expressions for the robust reliability, $\widehat{\alpha}$, as a function of the physical properties of the fin, and used these expressions to choose some of those properties, in particular the width and thickness profiles. The same relations can be employed to evaluate the sensitivity of the robustness to parameters which may be subject to fluctuation.

Consider for instance the local robustness in eq.(3.70) with constant width and thickness. $\widehat{\alpha}(v)$ has a minimum at the root of the fin, which determines the overall robustness:

$$\widehat{\alpha} = \frac{4WT^2 s_{cr}}{3L^2\sin^2\theta} \tag{3.82}$$

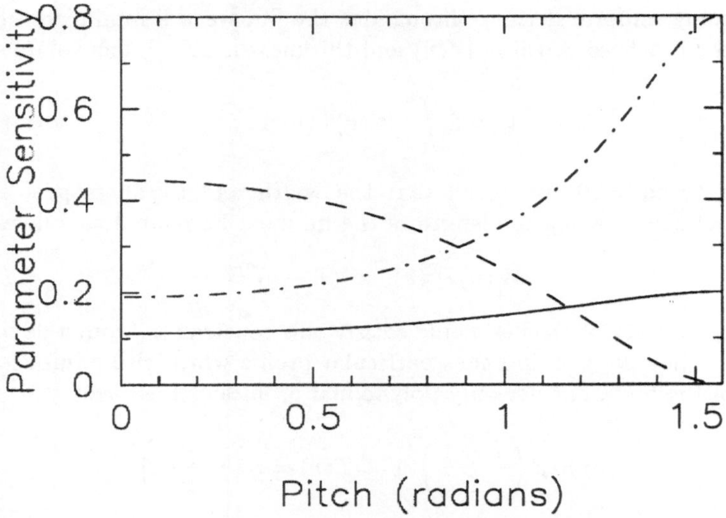

Figure 3.3: Normalized sensitivities of the robustness to the parameters s_{cr} (solid), θ (dash) and L (dash-dot).

Let us suppose that the width and thickness dimensions W and T are fairly immune to uncertain fluctuation. We will examine the sensitivity of $\widehat{\alpha}$ to the critical stress, s_{cr}, which is a material property, and to pitch θ, and length L which are geometrical. A convenient dimensionless measurement of sensitivity is the differential variation of $\widehat{\alpha}$ to the logarithm of the parameter:

$$\frac{\partial \widehat{\alpha}}{\partial \ln s_{cr}} = s_{cr} \frac{\partial \widehat{\alpha}}{\partial s_{cr}} = \widehat{\alpha} \tag{3.83}$$

$$\frac{\partial \widehat{\alpha}}{\partial \ln \theta} = \theta \frac{\partial \widehat{\alpha}}{\partial \theta} = -\frac{2\theta \widehat{\alpha}}{\tan \theta} \tag{3.84}$$

$$\frac{\partial \widehat{\alpha}}{\partial \ln L} = L \frac{\partial \widehat{\alpha}}{\partial L} = -2\widehat{\alpha} \tag{3.85}$$

Let $p^T = (s_{cr}, \theta, L)$ denote the parameter vector and let $\|\cdot\|$ denote the euclidean vector norm. It is useful to compare each individual squared sensitivity to the sum of the squares:

$$\frac{(\partial \widehat{\alpha}/\partial \ln s_{cr})^2}{\|\partial \widehat{\alpha}/\partial \ln p\|^2} = \frac{1}{5 + (2\theta/\tan\theta)^2} \tag{3.86}$$

$$\frac{(\partial \widehat{\alpha}/\partial \ln \theta)^2}{\|\partial \widehat{\alpha}/\partial \ln p\|^2} = \frac{1}{1 + 5(\tan\theta/2\theta)^2} \tag{3.87}$$

$$\frac{(\partial \widehat{\alpha}/\partial \ln L)^2}{\|\partial \widehat{\alpha}/\partial \ln p\|^2} = \frac{1}{(5/4) + (2\theta/\tan\theta)^2} \tag{3.88}$$

These relations express the relative sensitivity of the robustness parameter, $\widehat{\alpha}$, to the various physical parameters of the system. Fig. 3.3 shows the variation of these normalized squared sensitivities versus the pitch angle of the fin. At low angles of pitch it is the pitch itself to which the reliability is most sensitive, while at large pitch angle the beam length L has the greatest impact on the reliability.

3.4 Beam in Compression With Uncertain Initial Imperfections

An ideal, straight, uniform beam which is loaded in axial compression will buckle laterally only if the load exceeds a threshold value, called the Euler buckling load. However, if the beam has geometrical imperfections, then buckling can occur for loads below the Euler threshold. In this section we will consider the reliability of axially compressed beams with imperfect and uncertain shapes.

3.4.1 Band-Limited Energy-Bound Convex Model

In chapter 1 we defined an energy-bound convex model in which the uncertainty parameter, α, is the maximum strain-energy of geometric imperfections. This is a fruitful concept for robust reliability analysis, since the strain energy is also related to a criterion for failure, as we will see in this section. This will then lead directly to an expression for the robust reliability.[1]

We can imagine a set of initially straight beams having been warped to various shapes by the investment of a certain amount of energy. The amount of energy imposes a constraint on the shapes which can occur, but does not uniquely determine the shape of the warped beam. Conversely, we can imagine a set of warped beams requiring a certain amount of elastic energy to straighten them out. Consider a uniform beam of length L and flexural rigidity EI. Instead of the nominally straight equilibrium shape, consider the set $\mathcal{Y}(\alpha)$ of all shapes $y(x)$, consistent with the boundary conditions, whose energy of deformation does not exceed a constant value, α:

$$\mathcal{Y}(\alpha) = \left\{ y(x) : \ \frac{EI}{2} \int_0^L (\ddot{y}(x))^2 \, \mathrm{d}x \leq \alpha \right\} \tag{3.89}$$

Dots imply differentiation with respect to position. The unique nominal equilibrium shape, $y(x) = 0$, is replaced by an infinite set $\mathcal{Y}(\alpha)$ of uncertain beam profiles. The set $\mathcal{Y}(\alpha)$ is an *energy-bound convex model* for the uncertainty in the beam shape. We can think of $\mathcal{Y}(\alpha)$ as the set of all beams which have suffered an initial deformation consuming elastic energy no greater than α,

[1] This section is based in part on [12].

due to various environmental or manufacturing conditions or due to the operational history of the beam. It is implicitly assumed that the elements of a convex model such as $\mathcal{Y}(\alpha)$ satisfy all the boundary conditions inherent in the mechanical system.

It is useful to allow for the possibility that only a given range of mode shapes is involved in absorbing the initial deformation energy. Let us then define the ***band-limited energy-bound convex model*** as the set of functions of bounded deformation energy formed as the superposition of a finite number of sine functions. That is:

$$\mathcal{Y}(\alpha, N_0, N_1) = \left\{ y(x) = \sum_{n=N_0}^{N_1} a_n \sin \frac{n\pi x}{L} : \frac{EI}{2} \int_0^L (\ddot{y}(x))^2 \, dx \leq \alpha \right\} \tag{3.90}$$

This is the band-limited energy-bound convex model for representing uncertainty in the initial deformation of the beam.

3.4.2 Fourier Representation of $\mathcal{Y}(\alpha, N_0, N_1)$

Let the initial deformation of the beam be denoted $y_0(x)$ and expanded in a band-limited Fourier sine series on the interval $[0, L]$:

$$y_0(x) = \sum_{n=N_0}^{N_1} a_n \sin \frac{n\pi x}{L} \tag{3.91}$$

$y_0(x)$ is an arbitrary element of $\mathcal{Y}(\alpha, N_0, N_1)$, so it must satisfy:

$$\frac{EI}{2} \int_0^L (\ddot{y}(x))^2 \, dx \leq \alpha \tag{3.92}$$

Substituting (3.91) into (3.92) and employing the orthogonality of the sine functions, this becomes the following constraint on the Fourier coefficients $a_{N_0}, \ldots, a_{N_1}$:

$$\sum_{n=N_0}^{N_1} n^4 a_n^2 \leq \frac{4\alpha L}{\pi^2 P_E} \tag{3.93}$$

where $P_E = EI\pi^2/L^2$ is the critical Euler load. Thus $\mathcal{Y}(\alpha, N_0, N_1)$ can be expressed as the set of Fourier coefficients $a_{N_0}, \ldots, a_{N_1}$ which satisfy (3.93):

$$\mathcal{Y}(\alpha, N_0, N_1) = \left\{ a_{N_0}, \ldots, a_{N_1} : \sum_{n=N_0}^{N_1} n^4 a_n^2 \leq \frac{4\alpha L}{\pi^2 P_E} \right\} \tag{3.94}$$

This relation shows very clearly that $\mathcal{Y}(\alpha, N_0, N_1)$ is a convex set, whose Fourier representation is a multi-dimensional solid ellipsoid.

$\mathcal{Y}(\alpha, N_0, N_1)$ is the convex hull of the following set of boundary points, which constitute a multi-dimensional ellipsoidal shell:

$$\mathcal{B}(\alpha, N_0, N_1) = \left\{ a_{N_0}, \ldots, a_{N_1} : \sum_{n=N_0}^{N_1} n^4 a_n^2 = \frac{4\alpha L}{\pi^2 P_E} \right\} \tag{3.95}$$

3.4.3 Maximum Additional Bending Moment

We have defined the band-limited energy-bound convex model of uncertainty of the initial beam profile. Now we study the loaded beam and develop a failure criterion which accounts for the initial shape uncertainty.

We consider beams with stress-free but unknown initial deformations of bounded energy. Because the initial deflection is free of internal stresses, failure (according to the maximum-moment criterion) of the loaded beam results from the additional bending moments induced by the load.

Let us suppose that the initial deformation $y_0(x)$ belongs to the convex model $\mathcal{Y}(\alpha, N_0, N_1)$, and load the beam axially at each end with a compressive force P. The additional deformation is denoted $y_1(x)$ and satisfies the following differential equation:

$$EI\ddot{y}_1(x) = -P(y_0(x) + y_1(x)) \tag{3.96}$$

Denote the load ratio by $\rho = P/P_E$ and assume that $\rho < N_0$. The solution of eq.(3.96) is [98, p.33]:

$$y_1(x) = \rho \sum_{n=N_0}^{N_1} \frac{a_n}{n^2 - \rho} \sin \frac{n\pi x}{L} \tag{3.97}$$

where $a_{N_0}, \ldots, a_{N_1}$ are the Fourier coefficients of the initial deformation, $y_0(x)$. This relation shows that the initial imperfection enables deformation at loads below the Euler buckling load.

A common failure criterion is to presume that the beam fails if the bending stress in a tensile fiber exceeds the yield stress of the material. Since the initial deformation is stress-free, we calculate the maximum *additional* bending moment, $M_1 = EI\ddot{y}_1$. Employing eq.(3.97) to express the additional deflection at position ξ, the bending moment at position ξ becomes:

$$M_1(\xi) = -\rho P_E \sum_{n=N_0}^{N_1} \frac{n^2 a_n}{n^2 - \rho} \sin \frac{n\pi\xi}{L} \tag{3.98}$$

Thus, to establish the bending-moment criterion for failure, we must find the following extremum:

$$\max_{a_{N_0}, \ldots, a_{N_1}} \left[-\rho P_E \sum_{n=N_0}^{N_1} \frac{n^2 a_n}{n^2 - \rho} \sin \frac{n\pi\xi}{L} \right] \tag{3.99}$$

subject to the constraint implied by $\mathcal{Y}(\alpha, N_0, N_1)$, eq.(3.94). We replace the inequality in (3.94) by an equality because (3.99) is linear in the Fourier coefficients, $a_{N_0}, \ldots, a_{N_1}$. Using Lagrange optimization we obtain the following expression for the Fourier coefficients which optimize the magnitude of the bending moment at position ξ:

$$\widehat{a}_n(\xi) = \pm\sqrt{\frac{4\alpha L}{\pi^2 P_E}} \left(\sum_{n=N_0}^{N_1} \frac{\sin^2 n\pi\xi/L}{(n^2-\rho)^2} \right)^{-1/2} \frac{\sin n\pi\xi/L}{n^2(n^2-\rho)} \tag{3.100}$$

The $\pm$ is applied uniformly to all the coefficients, one choice producing the maximum and the other the minimum. The additional bending moment profile $M_1(x)$ of a beam with maximum moment at position ξ is:

$$\widehat{M}_1(x;\xi) = -\rho P_E \sum_{n=N_0}^{N_1} \frac{n^2 \widehat{a}_n(\xi)}{n^2-\rho} \sin n\pi x/L \tag{3.101}$$

To find the envelope of extremal additional bending moments we set $x = \xi$ in eq.(3.101):

$$\widehat{M}_1(\xi) = \pm\sqrt{\frac{4\rho^2 \alpha L P_E}{\pi^2} \sum_{n=N_0}^{N_1} \frac{\sin^2 n\pi\xi/L}{(n^2-\rho)^2}} \tag{3.102}$$

$\widehat{M}_1(\xi)$ is the greatest additional bending moment attainable at position ξ on the beam from any initial imperfection in $\mathcal{Y}(\alpha, N_0, N_1)$.

3.4.4 Critical-Energy Failure Criterion

We are now prepared to relate the energy of the initial uncertainty to the failure of the loaded beam, and then to evaluate the robust reliability of the beam. The maximum-stress or maximum-moment concepts are local failure criteria, but they are governed by global energetic considerations. A beam fails locally, according to the maximum stress concept, when the bending stress reaches the yield value. From the energetic point of view, however, this local yielding can only occur when sufficient work has been done on the entire beam to bring the normal stress on a beam section at one point to its critical value.

In the absence of initial deflections, the axially compressed beam allows no equilibrium deflection for loads less than the critical Euler load. However, the initial deformation of the beam allows the Euler buckling modes to appear even for $P < P_E$, as shown by eq.(3.97). Furthermore, the strain energy of the initial elastic deformation, α, controls the maximum bending moment which can be achieved (eq.(3.102)). Following this idea, we can define a critical value of the initial deformation energy, α, beyond which failure can occur for some beams in $\mathcal{Y}(\alpha, N_0, N_1)$ when loaded to the load ratio ρ, but below which it cannot. This is useful because the initial deformation energy also defines the degree of uncertainty in the initial beam shape, α in eq. (3.90). Thus the failure criterion is directly related, through a single physical parameter, to the initial uncertainty. This critical α is precisely the robust reliability: the greatest uncertainty which the beam can tolerate without failing.

Let σ_y be the tensile yield stress of the beam material, let c be the distance from the neutral plane of the beam to the most distant tensile fiber, and let I be the moment of inertia of the beam cross section with respect to the neutral plane. The maximum moment concept states that the beam fails if the maximum additional bending moment reaches the critical value:

$$M_{\rm cr} = \frac{\sigma_y I}{c} \tag{3.103}$$

The maximum additional bending moment is obtained from eq.(3.102) by maximizing on the position variable:

$$M_{\max} = \sqrt{\frac{4\rho^2 \alpha L P_E}{\pi^2} \max_{\xi} \sum_{n=N_0}^{N_1} \frac{\sin^2 n\pi\xi/L}{(n^2-\rho)^2}} \tag{3.104}$$

Equating eqs.(3.103) and (3.104) and solving for α, we obtain the critical value of the initial deformation energy, beyond which beam failure will occur for some of the ensemble of beams generated by the uncertainty in the initial deformation, which is precisely the robust reliability:

$$\widehat{\alpha} = \frac{\pi^2 \sigma_y^2 I^2}{4c^2 \rho^2 L P_E} \left(\max_{\xi} \sum_{n=N_0}^{N_1} \frac{\sin^2 n\pi\xi/L}{(n^2-\rho)^2} \right)^{-1} \tag{3.105}$$

Not every beam whose initial deflection energy equals $\widehat{\alpha}$ will fail when loaded to a load-ratio ρ. However at least one such beam will fail by having a local bending stress equal to σ_y. Conversely, if α' is less than $\widehat{\alpha}$, then no beam will fail whose initial deformation is in $\mathcal{Y}(\alpha', N_0, N_1)$ and is then compressed to a load ratio ρ. The deformation energy in eq.(3.105) thus defines the largest set $\mathcal{Y}(\alpha, N_0, N_1)$ of initial deformations for which the resulting loaded beams will have maximum bending stress not exceeding σ_y. $\widehat{\alpha}$ is the amount of initial imperfection-uncertainty, in units of energy, which the system can tolerate. When $\widehat{\alpha}$ is large the beam is robust with respect to imperfections; when $\widehat{\alpha}$ is small the beam is fragile to imperfection.

Eq.(3.105) can be used to plot ρ versus α. Fig. 3.4 shows the dependence of the critical load ratio, ρ, on the initial deformation energy α, for several values of the initial deflection mode numbers N_0 and N_1. The critical energy is in units of $U_e = \pi^2 {\sigma_y}^2 I^2 / 4c^2 L P_E$, which is the strain energy of an Euler beam in buckling. These curves illustrate that the critical load ratio is very sensitive to the initial deformation energy. ρ asymptotically approaches zero as α increases. Conversely, ρ converges to N_0^2 as the initial deformation energy vanishes. Furthermore, the sum in eq.(3.105) converges rapidly and

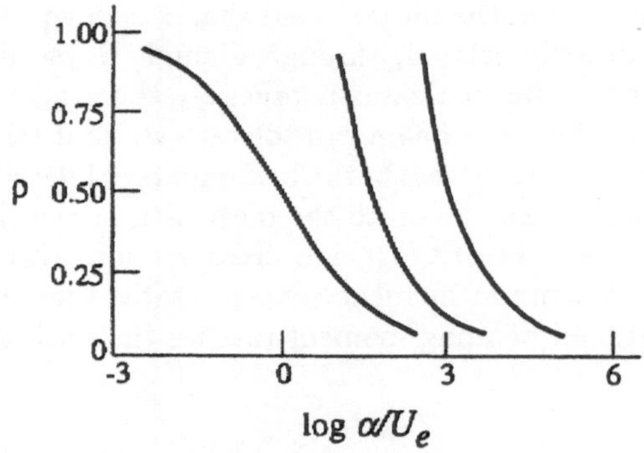

Figure 3.4: Critical load ratio versus initial deformation energy. N_1, N_2 = 1, 10 (left); 2,11 (middle); 5, 15 (right).

is dominated by the lowest terms. Low-order modes are much more sensitive to initial deflection energy than high-order modes.

If $N_0 = 1$ the maximum in eq.(3.105) occurs near $\xi = L/2$, and the sum can be truncated after the first term without too much error. One then finds that the load ratio, ρ, and the initial deformation energy, α, satisfy the following approximate relationship at the onset of yielding:

$$\left(\frac{\rho}{1-\rho}\right)^2 \widehat{\alpha} \approx \frac{\pi^2 \sigma_y^2 I^2}{4c^2 L P_E} \tag{3.106}$$

The righthand side of eq.(3.106) expresses the geometrical and material properties of the beam, while the lefthand side involves the load ratio ρ and the robustness $\widehat{\alpha}$. The relationship expresses the trade-off between the load and the reliability. The reliability $\widehat{\alpha}$ decreases from infinity to zero as the load ratio ρ rises from zero to unity.

3.5 Radial Buckling of Thin-Walled Shells: Reliability and Quality Control

The load-bearing capability of thin-walled structures is greatly reduced by the presence of even small geometrical imperfections. From the point of view of robust reliability this means that these structures are fragile with respect to uncertainty in their geometry. In this section we evaluate the robust reliability of a shell with uncertain localized imperfections, and subject to radial pulse loading. In section 3.5.1 we characterize the uncertain imperfections in terms of the radial tolerance of the shell by using a uniform-bound convex model. In determining the reliability of the shell in terms of the radial tolerance to which it is manufactured, we establish a connection between

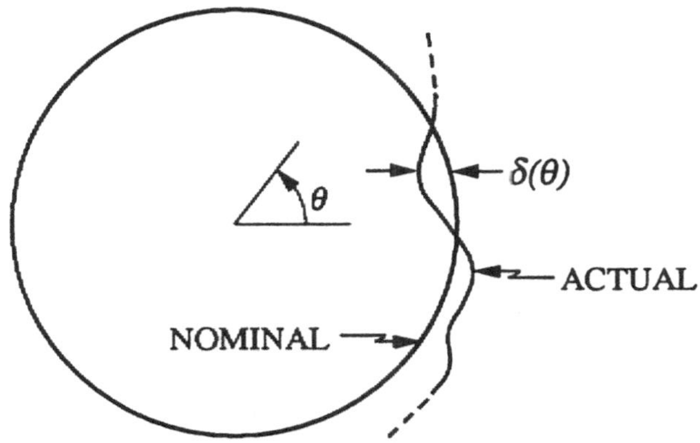

Figure 3.5: Cross section of a thin-walled shell showing actual and nominal shapes.

reliability and a simple quality assurance parameter. In section 3.5.2 we represent the uncertainty with an Fourier ellipsoid-bound convex model.

3.5.1 Localized Imperfections

Consider a cross section of the shell before loading, as in fig. 3.5. The actual shape deviates from the nominal circular shape due to geometrical imperfections which are of course greatly exaggerated in this figure.

We will explore the dependence of the shell reliability on the amplitude and angular size of the geometrical imperfections. The initial deviation of the shell from its nominal shape, as a function of azimuthal position θ, is represented by the imperfection function $\delta(\theta)$. We wish to represent uncertainty in imperfections which subtend ψ radians and have a maximum amplitude α. Consequently, we use the local uniform-bound convex model:

$$\mathcal{D}_{\rm lub}(\alpha,\psi) = \{\delta(\theta) : \ |\delta(\theta)| \le \alpha\chi(\psi-\theta)\} \tag{3.107}$$

where the characteristic function $\chi(\phi)$ is defined as:

$$\chi(\phi) = \begin{cases} 0, & \phi < 0 \\ 1, & \phi \ge 0 \end{cases} \tag{3.108}$$

The imperfections in the set $\mathcal{D}_{\rm lub}$ subtend an angle of ψ degrees, and the amplitude of the imperfections varies between $-\alpha$ and $+\alpha$.

We will discuss a method for describing pulsed buckling phenomena which was derived by Abrahamson and Goodier [1] and is extensively discussed in Lindberg and Florence [60]. Linear deflection equations are obtained by treating the material as perfectly plastic, by relating the initial geometrical imperfections to the initial radial velocity of the shell, and by considering only short times after application of the load. The additional deflection of the shell at angle θ and time τ after loading is [61, eq.(14)]:

$$u(\theta,\tau) = \int_0^{2\pi} \delta(\xi) S(\xi,\theta,\tau)\, d\xi \tag{3.109}$$

where

$$S(\xi,\theta,\tau) = \frac{1}{\pi} \sum_{n=2}^{N} G_n(\tau) \cos n(\theta - \xi) \tag{3.110}$$

The value of N is chosen to include all of the functions G_n of significant amplitude. The functions $G_n(\tau)$ are called "amplification functions" since G_n modulates the contribution of the nth spatial frequency. G_n is defined as:

$$G_n(\tau) = \begin{cases} \frac{1}{\eta^2-1}(1 - \cosh p_n\tau) & \eta < 1 \\ \frac{1}{\eta^2-1}(1 - \cos p_n\tau) & \eta > 1 \end{cases} \tag{3.111}$$

where

$$\eta = n/s \quad , \quad p_n = \sqrt{|[\eta^2 - (1/s^2)][1-\eta^2]|} \quad , \quad \tau = \frac{s^2 c_p h}{a^2\sqrt{12}} t \tag{3.112}$$

and

$$s^2 = \frac{12 a^2 \sigma_y}{h^2 E_h} \quad , \quad c_p = \sqrt{\frac{E_h}{\rho}} \tag{3.113}$$

where h is the wall thickness, a is the shell radius, σ_y is the yield stress, E_h is the strain hardening modulus and ρ is the density. The amplification function for $\eta = 1$ is $G_n(\tau) = \tau^2/2$.

The maximum response for $\delta(\xi) \in \mathcal{D}_{lub}(\alpha,\psi)$ is obtained when $\delta(\xi)$ switches back and forth between $+\alpha$ and $-\alpha$ as $S(\xi,\theta,\tau)$ changes sign. The maximum response at angle θ becomes:

$$u_{max}(\theta,\tau) = \max_{\delta\in\mathcal{D}_{lub}} u(\theta,\tau) = \alpha \int_0^{\psi} |S(\xi,\theta,\tau)|\, d\xi \tag{3.114}$$

From eq.(3.114), and exploiting the rotational symmetry of the function $S(\xi,\theta,\tau)$, we can express the maximum deflection at time τ, which occurs at the midpoint of the imperfected zone, as:

$$u_{max}(\tau) = u_{max}(\psi/2,\tau) = \alpha \int_{-\psi/2}^{\psi/2} |S(\xi,0,\tau)|\, d\xi \tag{3.115}$$

Values of u_{max}/α are presented in table 3.1, for parameter values $s = 20$, $\tau = 6$ and $N = 50$.

Excessive deflection of the shell a short time after loading is an indication of subsequent failure. Experience in this field has led to the following

ψ (deg)	$u_{max}(\tau)/\alpha$
1	1.38
3	4.00
5	6.25
7	7.94
10	9.21
15	10.6
20	14.6
30	21.9
50	27.4
120	31.2
360	31.8

Table 3.1: The maximum response based on the localized uniform bound convex model.

deflection-based failure criterion. Failure occurs when the deflection at time τ exceeds the critical value u_{cr}:

$$u_{max} > u_{cr} \tag{3.116}$$

Combining eqs.(3.115) and (3.116) determines the greatest amplitude of imperfection uncertainty which the shell can tolerate:

$$\widehat{\alpha}(\psi) = \frac{u_{cr}}{\int_{-\psi/2}^{\psi/2} |S(\xi, 0, \tau)| \, d\xi} \tag{3.117}$$

When $\widehat{\alpha}$ is large the shell is robust to geometrical imperfections; when $\widehat{\alpha}$ is small the shell is fragile to even small deviations from its nominal shape. $\widehat{\alpha}$ expresses the robustness of the shell to the *amplitude* of geometrical imperfections.

In eq.(3.117) we have expressed the limiting imperfection amplitude $\widehat{\alpha}$ as a function of the angular width ψ of the imperfected region of the shell. We could just as well use eq.(3.117) to express the greatest tolerable angular width of an imperfected region, as a function of the maximum imperfection amplitude, $\widehat{\psi}(\alpha)$. When $\widehat{\psi}$ is large the shell is robust to the angular size of imperfections, while a small value of $\widehat{\psi}$ indicates that the shell is sensitive to even small regions of geometrical imperfection. $\widehat{\psi}(\alpha)$ expresses the robustness of the shell to the *angular width* of geometrical imperfections.

More generally, we could plot a curve of constant $u_{max} = u_{cr}$ on ψ-vs-α coordinates. Points (α, ψ) below the curve represent convex models of imperfections which are robust, which will not fail, according to the criterion of eq.(3.116). Points above the curve represent convex models which include imperfections which will violate the condition for no-failure. The border between these two regions is the *robustness curve*. The results of table 3.1 are used to construct such a plot, for $u_{cr} = 1$, as shown in fig. 3.6.

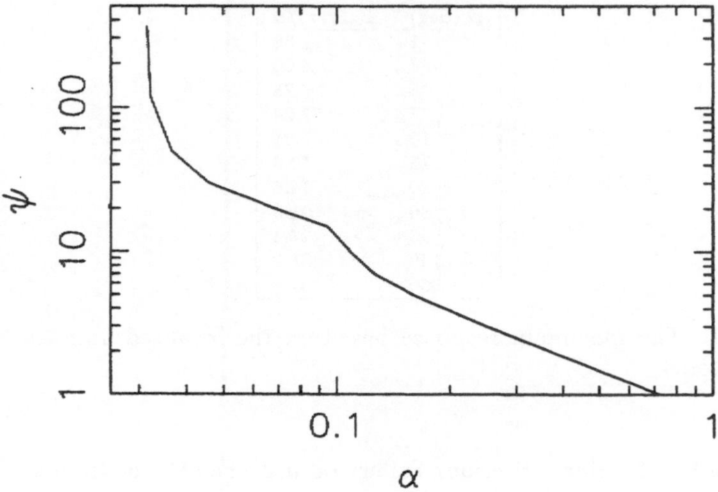

Figure 3.6: Robustness curve.

We can use fig. 3.6 as a manufacturing and inspection guideline for quality assurance. If we inspect for imperfections subtending an angle no smaller than ψ, then the curve establishes the greatest acceptable radial tolerance α of these imperfections.

3.5.2 Fourier Ellipsoid-Bound

The local uniform bound convex model presumes that our information about the uncertain imperfections is limited to radial tolerance and angular range. The reliability estimate may be substantially improved if spectral information about the imperfection is also available.

The spectral information about the imperfections constrains the spatial frequencies which contribute to the geometrical deviation of the shell from its nominal shape. Rather than allowing the imperfection profile to assume any spatial frequency, the initial deviation of the shell from its nominal shape is described by a truncated Fourier series. In other words, the initial imperfections are represented as the sum of certain spatial modes:

$$\delta(\theta) = \sum_{n=2}^{N} (a_n \cos n\theta + b_n \sin n\theta) \tag{3.118}$$

Let D be the vector of Fourier coefficients:

$$D^T = (a_2, a_3, \ldots, a_N, b_2, b_3, \ldots, b_N) \tag{3.119}$$

Uncertainty in these imperfections is described by the Fourier ellipsoid-bound convex model:

$$\mathcal{D}_{\text{feb}} = \left\{ D: \ D^T W D \le \alpha^2 \right\} \tag{3.120}$$

The shape of the ellipsoid determines the greatest possible contributions of the different spatial modes to the initial imperfections. The shape of the ellipsoid is determined by the positive definite real symmetric matrix W.

The deformation of the shell at normalized time τ after the impulse loading and at azimuthal position θ is also expressed as a sum of spatial modal components:

$$u(\theta,\tau) = \sum_{n=2}^{N} \left[a_n G_n(\tau) \cos n\theta + b_n G_n(\tau) \sin n\theta \right] \tag{3.121}$$

$$= D^T \phi(\theta,\tau) \tag{3.122}$$

where

$$\phi(\theta,\tau)^T = (G_2(\tau)\cos 2\theta,\ G_3(\tau)\cos 3\theta,\ \ldots,\ G_N(\tau)\cos N\theta,\ G_2(\tau)\sin 2\theta,\ G_3(\tau)\sin 3\theta,\ \ldots,\ G_N(\tau)\sin N\theta) \tag{3.123}$$

The importance of a mode is determined by the corresponding amplification function G_n, defined in eq.(3.111). When G_n is large then that spatial mode will emerge prominently in the response. The distinctive feature of the ellipsoid-bound convex model is that it allows one to restrict the contribution to the initial imperfections of modes which are dominant in the response. Of course, one cannot arbitrarily restrict the modal components of the initial imperfections; this must be based on information about these imperfections. But here is precisely the point of contact with quality control considerations in manufacturing. Our present analysis allows us to determine which spatial modes of imperfection, if strictly controlled in manufacture, will enhance the immunity of the shells to initial imperfections.

The greatest deflection at angle θ, in response to a radial impulse load, is the maximum of $u(\theta,\tau)$ on the set $\mathcal{D}_{\text{feb}}$ of allowed initial imperfection profiles:

$$u_{\max}(\theta,\tau) = \max_{D \in \mathcal{D}_{\text{feb}}} D^T \phi(\theta,\tau) \tag{3.124}$$

This is the maximum of a linear function, $D^T\phi(\theta,\tau)$, on a convex set, $\mathcal{D}_{\text{feb}}$, and thus occurs on the boundary of the set. The method of Lagrange multipliers again provides an immediate solution [61]:

$$u_{\max}(\theta,\tau) = \alpha\sqrt{\phi(\theta,\tau)^T W^{-1} \phi(\theta,\tau)} \tag{3.125}$$

Combining this relation with the condition for failure, eq.(3.116), and maximizing on the angle, leads to an expression for the greatest uncertainty, $\widehat{\alpha}$, consistent with no failure of the shell:

$$\widehat{\alpha} = \frac{u_{\text{cr}}}{\sqrt{\max_\theta \phi(\theta,\tau)^T W^{-1} \phi(\theta,\tau)}} \tag{3.126}$$

Eq.(3.126) is the robust reliability of the shell in terms of the spectral uncertainty in the imperfections, as represented by $\mathcal{D}_{\text{feb}}$. When $\widehat{\alpha}$ is large then the shells can tolerate large uncertainty, while a small value of $\widehat{\alpha}$ implies that the shells are fragile with respect to the imperfections.

Let us now consider the two measures of reliability which we have derived, eqs.(3.117) and (3.126). The former expresses the robust reliability in terms of the radial tolerance to which the shell was produced. It is based on the local uniform-bound convex model $\mathcal{D}_{\text{lub}}$, and we denote it α_{lub}. Eq.(3.126) is the robust reliability of the shell in terms of spectral uncertainty, it is derived from the Fourier ellipsoid-bound convex model $\mathcal{D}_{\text{feb}}$, and will be denoted α_{feb}. These two measures of reliability are not directly comparable since they are derived from distinct uncertainty models and their units differ.

However, one can determine the greatest initial imperfection for any shell in the ellipsoid-bound convex model (problem 4). This quantity, call it $\widehat{\delta}$, can be directly compared with α_{lub}, which also has units of imperfection size. The ellipsoid-bound uncertainty model $\mathcal{D}_{\text{feb}}$ expresses the relative importance of the various buckling modes; some modes contribute more and some less to the initial imperfections. Likewise the response to impulse loading, eq.(3.121), expresses the importance of the different buckling modes to the response. If $\mathcal{D}_{\text{feb}}$ strongly constrains the amplitudes of modes which dominate the response, then the ellipsoidal model will allow large initial imperfections, and thus will result in greater robust reliability of the shell. This is explored further in problem 5.

3.6 Reliability of Serial and Parallel Networks

In the previous sections we have considered the robust reliability of single components or structures. We now consider the reliability analysis of a network of interconnected units. We will establish the reliability of the network in terms of the subunit reliabilities, by assuming that the subunits fail or survive independently of one another, and that each subunit is subject to the same source of uncertainty.

For instance, we may consider a collection of N cooling fins like the one studied in section 3.3. These fins may be different from each other, each with its own length, pitch and critical bending moment. The uncertainty associated with the nth subunit is represented by the convex model $\mathcal{U}_n(\alpha)$, and the robust reliability of the nth unit is α_n. While each subunit has its own mechanical model, failure criterion and uncertainty model, the uncertainty parameter α is common to all the subunits.

Suppose that the cooling network of N fins fails if any single fin fails. The greatest uncertainty which the network can tolerate is equal to the greatest uncertainty which the least reliable subunit can tolerate: as soon as the uncertainty increases to the point where failure becomes possible for the least reliable subsystem, the entire network is vulnerable to failure. In this

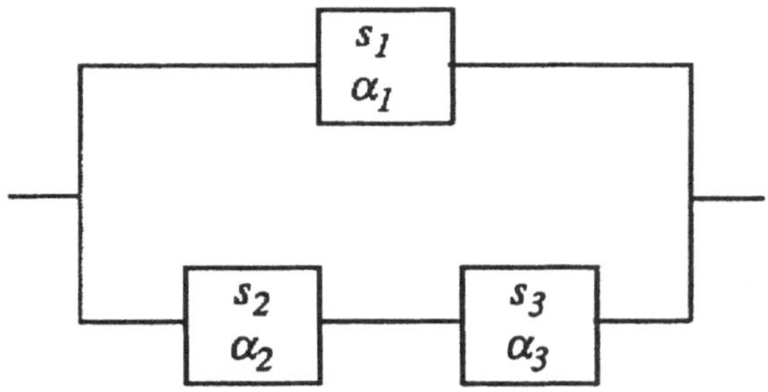

Figure 3.7: A network with three independent subunits.

situation the network reliability equals the least value from among the subunit reliabilities:

$$\widehat{\alpha} = \min_{1\leq n\leq N} \alpha_n \tag{3.127}$$

This relation describes the reliability of N *independent serial subunits*, in analogy to a network of serial electronic ciruits which fails if any single circuit fails.

At the other extreme, suppose the network continues to operate as long as even a single subunit remains operational. It is now the most reliable subunit which controls the network reliability:

$$\widehat{\alpha} = \max_{1\leq n\leq N} \alpha_n \tag{3.128}$$

Again using the electronic analogy, this is the reliability of N *independent parallel subunits*.

Let us now consider the general situation, in which the network fails when n out of the N subunits fail. Let us arrange the subunit reliabilities in increasing order:

$$\alpha_{r_1} \leq \alpha_{r_2} \leq \cdots \leq \alpha_{r_N} \tag{3.129}$$

Subunit r_1 is the least reliable, becoming eligible for failure when the uncertainty parameter is no less than α_{r_1}. Similarly, subunit r_2 is the next "weakest" subunit, while subunit r_N is the most reliable. Failure of n subunits becomes possible if the uncertainty parameter is at least as large as α_{r_n}, so the network reliability is:

$$\widehat{\alpha} = \alpha_{r_n} \tag{3.130}$$

This relation includes eqs.(3.127) and (3.128) as special cases.

Example 3 Consider the network shown in fig. 3.7, with three independent subunits subject to a common source of uncertainty. The individual subunit reliabilities are shown in the figure. The network as a whole remains operational if either the upper branch (S_1) or the lower branch (S_2 and S_3) are

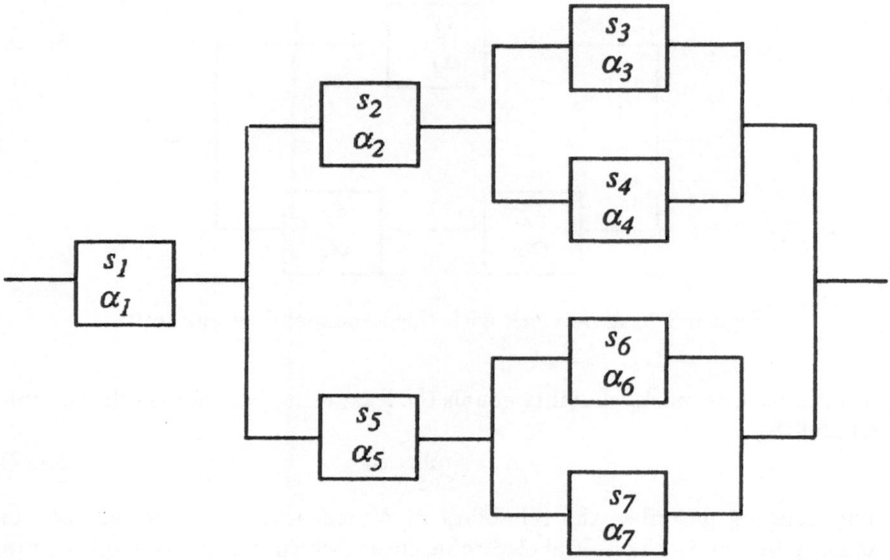

Figure 3.8: A network with seven independent subunits.

functional. Following the line of reasoning we have developed, we conclude that the network reliability is:

$$\widehat{\alpha} = \max\,[\alpha_1,\, \min(\alpha_2,\, \alpha_3)] \tag{3.131}$$

The rationale for this relation is that S_2 is in series with S_3, and that S_1 is parallel to $(S_2,\, S_3)$. ■

Diagrams such as fig. 3.7 are useful for representing parallel and serial components of a network, and for indicating the min/max structure of the network reliability. Let us consider a more complex example.

Example 4 The network of fig. 3.8 has seven independent subunits. S_3 and S_4 are parallel, so the reliability of this couplet is:

$$\alpha_{3,4} = \max[\alpha_3, \alpha_4] \tag{3.132}$$

Subunit S_2 is in series with the parallel couple $(S_3,\, S_4)$, so the reliability of these three subunits is:

$$\alpha_{2,3,4} = \min[\alpha_2,\, \alpha_{3,4}] = \min[\alpha_2,\, \max(\alpha_3, \alpha_4)] \tag{3.133}$$

Likewise, $(S_6,\, S_7)$ is a parallel couplet in series with S_5, so the reliability of these three subunits is:

$$\alpha_{5,6,7} = \min[\alpha_5,\, \alpha_{6,7}] = \min[\alpha_5,\, \max(\alpha_6, \alpha_7)] \tag{3.134}$$

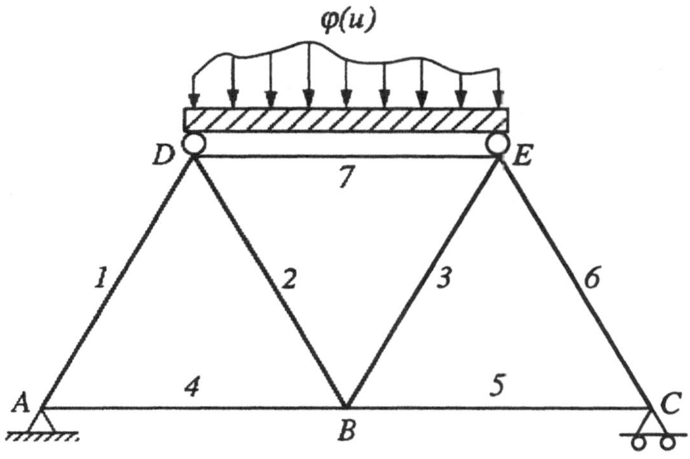

Figure 3.9: Truss with uncertain load for problem 1.

The triplets (S_2, S_3, S_4) and (S_5, S_6, S_7) are in parallel, so their reliability is:

$$\alpha_{2\ldots7} = \max[\alpha_{2,3,4},\ \alpha_{5,6,7}] \tag{3.135}$$

Finally, S_1 is parallel to (S_2, ..., S_7), so the reliability of the entire network is:

$$\widehat{\alpha} = \min[\alpha_1,\ \alpha_{2\ldots7}] \tag{3.136}$$

In this recursive manner we derive the reliability of a complex network of independent subunits subject to a common source of uncertainty. ■

3.7 Problems

1. The beams in the truss of fig. 3.9 are all of length L and have uniform cross section. The cross sectional area of the nth beam is A_n. The uncertainty in the distributed load $\phi(u)$ is described by the following convex model:

$$\mathcal{U}(\alpha) = \{\phi(u) :\ |\phi(u)| \le \alpha\} \tag{3.137}$$

   (a) Calculate the robust reliabilities of beams 2 and 4, which fail if the normal stress exceeds the σ_y. (b) Determine the ratio of areas of beams 2 and 4 so that their robust reliabilities are the same.

2. ‡ Develop a generalization of the minimum-weight uniform-reliability result of section 3.3.4. Show that the conclusion of section 3.3.4 is valid for arbitrary width and thickness functions. Hint: use the following functions to represent general width and thickness profiles which satisfy

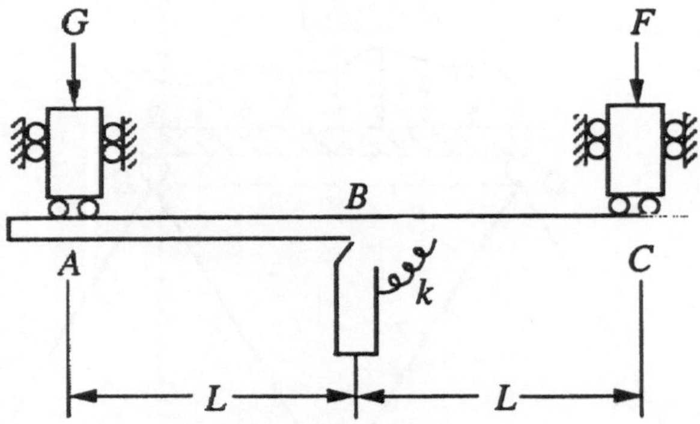

Figure 3.10: Structure for problem 6.

the uniform-reliability criterion (3.78):

$$W(v) = \beta \left(\frac{L-v}{L} \right)^{\gamma} f^2(v/L), \quad T(v) = \eta \left(\frac{L-v}{L} \right)^{\xi} \frac{1}{f(v/L)} \tag{3.138}$$

where $f(v/L)$ is an arbitrary positive function.

3. Use Lagrange optimization to derive relations (3.100) and (3.101).

4. ‡ Find the greatest amplitude of an initial imperfection in the Fourier ellipsoid-bound convex model, eq.(3.126).

5. (a) Perform numerical calculations to determine the relative importance of the modes in the response function, eq.(3.121). Assume the Fourier coefficients are all equal and use parameter values $s = 20$, $\tau = 6$ and $N = 50$. (b) Let the matrix W in the convex model $\mathcal{D}_{\text{feb}}$, eq.(3.120), be diagonal. Use the result of (a) to suggest a desirable W, with positive elements which sum to unity. What is the implication of this W for quality-control of the shell manufacturing process? ‡(c) What positive diagonal matrix, W, minimizes $u_{\max}$?

6. The structure in fig. 3.10 is a horizontal platform which is loaded by two static vertical forces, F and G and by a rotational spring of stiffness k. The spring applies a moment $M_s = k\theta$ when the bar AC is tilted at an angle θ. The forces F and G are uncertain and bounded:

$$|F| \le \alpha, \quad |G| \le \alpha \tag{3.139}$$

The platform is satisfactorily horizontal if the angle of tilt is no greater than a given critical value:

$$|\theta| \le \theta_{\text{cr}} \tag{3.140}$$

Determine the robust reliability of the platform.

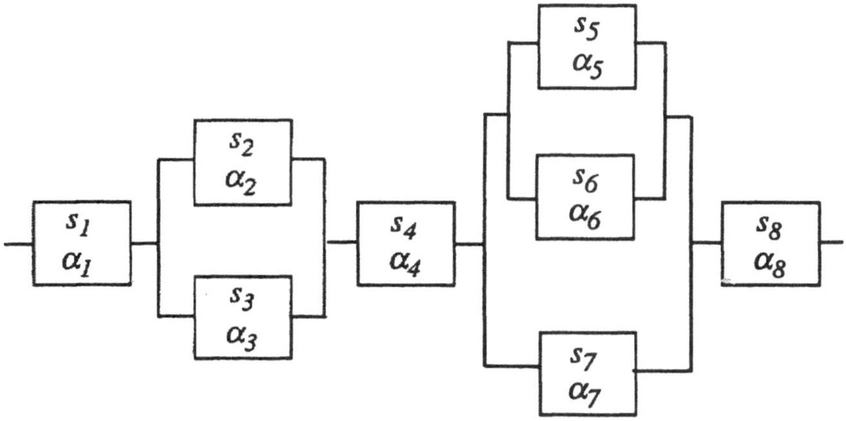

Figure 3.11: Network for problem 7.

7. Evaluate the robust reliability of the network of independent subunits in fig. 3.11.

8. (a) Consider the three-unit network of fig. 3.7. If we could choose the subunit reliabilities such that $\alpha_1 + \alpha_2 + \alpha_3 = 1$, what choice would maximize the network reliability? (b) Suppose we could invest a limited resource, q, to improve the reliability of each subunit, according to the relation $\alpha_n = \rho_n q_n$, where ρ_n is a constant. The total amount of the resource available is Q, where $q_1 + q_2 + q_3 = Q$. What allocation optimizes the network reliability?

9. *Cooling fin with uncertain heat transfer coefficient.* Fins are used to enhance the heat transfer between a solid body and a fluid. The local convective heat flux [Watts/m^2] at position x along a thin infinitely long fin in steady state is [77, p.569]:

$$q''(x) = h(T_0 - T_\infty)\mathrm{e}^{-mx\sqrt{h}} \tag{3.141}$$

where T_0 and T_∞ are root and tip temperatures, h is the heat transfer coefficient of the fin surface, and m is a constant. The fin is designed for heat flux not exceeding a critical value, so failure of the fin can be defined as:

$$\max_x q''(x) \geq q''_{\mathrm{cr}} \tag{3.142}$$

The heat transfer coefficient is constant over the fin surface, but subject to interval uncertainty:

$$\left|h - \overline{h}\right| \leq \alpha \tag{3.143}$$

What is the robust reliability of the fin?

Chapter 4

Robust Reliability of Time-Varying Systems

In chapter 3 we studied the robust reliability of primarily static systems with time-invariant uncertainties. In the present chapter we will consider time-varying uncertainties and dynamic systems or processes which evolve in time. In sections 4.1 to 4.3 we consider linear elastic vibrations driven by uncertain time-varying loads. Uncertainties may exist in both the inputs and the failure states of the structure, which will lead to the ideas of input reliability and failure reliability, as well as the overall reliability of the system. When we consider general multi-dimensional systems we will assess the relative reliability of the individual modes or degrees of freedom, which is developed in section 4.4. In section 4.5 we study the reliability of dynamic buckling of an axially loaded shell. In section 4.6 we develop the robust reliability of a dynamically loaded structure which is vulnerable to fatigue failure under uncertain repetitive loading.

4.1 Mass and Spring System

Consider a one-dimensional undamped linear oscillator driven by uncertain inputs $u(t)$:

$$m\ddot{x}(t) + kx(t) = u(t), \quad x(0) = \dot{x}(0) = 0 \tag{4.1}$$

The input $u(t)$ is an uncertain transient function described by a cumulative energy-bound convex model centered at the origin:

$$\mathcal{U}(\alpha_i) = \left\{ u(t) : \int_0^\infty u^2(t)\,\mathrm{d}t \le \alpha_i^2 \right\} \tag{4.2}$$

The uncertainty parameter for the input is α_i. The response to input $u(t)$ is:

$$x_u(t) = \frac{1}{m\omega} \int_0^t u(\tau) \sin \omega(t-\tau)\, \mathrm{d}\tau \tag{4.3}$$

where the natural frequency is $\omega = \sqrt{k/m}$.

Each input function produces a unique output function, and the input set as a whole generates a set of outputs, called the *response set*, which we denote:

$$\mathcal{X}(\alpha_i) = \{x_u(t) \ \text{ for all } \ u \in \mathcal{U}(\alpha_i)\} \tag{4.4}$$

We will suppose that failure occurs if the output is too close to a particular critical value, $\overline{f}$, and furthermore that this critical output value is itself uncertain. We will use a convex model to represent the uncertain failure values by defining the *failure set* as:

$$\mathcal{F}(\alpha_f) = \left\{ f(t) : \ \left[f(t) - \overline{f}\right]^2 < \alpha_f^2 \right\} \tag{4.5}$$

For any function $f(t) \in \mathcal{F}(\alpha_f)$, the system is considered to have failed at time t if $x(t) = f(t)$. The nominal failure state is $\overline{f}$, but any state which is within $\pm\alpha_f$ of $\overline{f}$ constitutes failure. The parameter α_f expresses the amount of uncertainty in the value of the failure state. For simplicity we assume that $0 < \alpha_f \leq \overline{f}$.

The robust reliability of a system is a measure of its immunity to uncertainty. Since we have uncertainty both in the input function and in the failure state, we can evaluate the robust reliability with respect to either of these uncertainties separately, or to both of them together. That is, we can ask each of the following questions.

1. For a given level α_f of failure uncertainty, how much input uncertainty α_i can the system tolerate without failing? We will call this quantity the *input reliability*, and denote it $\widehat{\alpha}_i$. Sometimes we will explicitly indicate the dependence of the input reliability on time and on the level of failure uncertainty by writing $\widehat{\alpha}_i(t, \alpha_f)$.

2. For a given level α_i of input uncertainty, how much failure uncertainty α_f can the system tolerate without failing? We will call this the *failure reliability*, and denote it $\widehat{\alpha}_f$ or, more completely, $\widehat{\alpha}_f(t, \alpha_i)$.

3. Considering input and failure uncertainties together, how much uncertainty in both of these quantities can the system tolerate? This is the *overall reliability*, represented by $\widehat{\alpha}$ or $\widehat{\alpha}(t)$.

The system fails when the response hits a failure state: $x_u(t) = f(t)$. In other words, failure can occur at time t if an element of the response set, $\mathcal{X}(\alpha_i)$, takes a value which also occurs at the same time in the failure set, $\mathcal{F}(\alpha_f)$. Consequently, in all three of these reliability analyses, evaluation

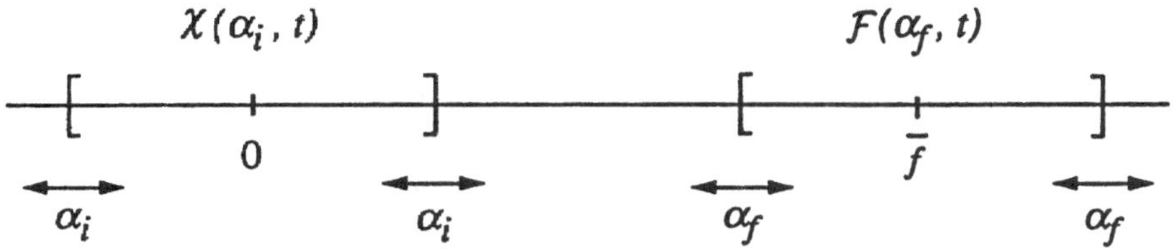

Figure 4.1: Response and failure sets at time t, showing their expansion and contraction with the uncertainty parameters α_i and α_f.

of the reliability at an instant t hinges upon determining the disjointness of response and failure sets. Failure does not occur if and only if $\mathcal{X}(\alpha_i)$ and $\mathcal{F}(\alpha_f)$ are disjoint:

$$\mathcal{X}(\alpha_i) \cap \mathcal{F}(\alpha_f) = \emptyset \tag{4.6}$$

As explained in section 2.3, the convex sets $\mathcal{X}(\alpha_i)$ and $\mathcal{F}(\alpha_f)$ expand and contract as their uncertainty parameters α_i and α_f grow and diminish. $\mathcal{X}(\alpha_i)$ and $\mathcal{F}(\alpha_f)$ are sets of scalar functions, but the values assumed by these functions at any instant simply define intervals, as shown in fig. 4.1. That is, fig. 4.1 is a cross-section of $\mathcal{X}(\alpha_i)$ and $\mathcal{F}(\alpha_f)$ at a particular instant in time. $\mathcal{X}(\alpha_i, t)$ is an interval containing the origin. This interval will be large or small, depending on the value of α_i. Similarly, $\mathcal{F}(\alpha_f, t)$ is an interval around $\overline{f}$, and the interval grows and shrinks with α_f.

The input reliability at time t is the upper bound of the values of α_i for which $\mathcal{X}(\alpha_i, t)$ and $\mathcal{F}(\alpha_f, t)$ are disjoint, for fixed failure uncertainty α_f. Similarly, the failure reliability is the upper bound α_f values for which $\mathcal{X}(\alpha_i, t)$ and $\mathcal{F}(\alpha_f, t)$ are disjoint, for fixed input uncertainty α_i. Finally, the overall reliability is obtained by treating α_i and α_f as the same uncertainty parameter, α, and seeking the upper bound of α values at which the sets are disjoint.

The input and response sets are centered at the origin, while the failure set is centered at the positive value $\overline{f}$, so the disjointness condition of relation (4.6) is equivalent to:

$$\max_{u \in \mathcal{U}(\alpha_i)} x_u(t) < \min_{f \in \mathcal{F}(\alpha_f)} f(t) \tag{4.7}$$

The minimum on the right is simply $\overline{f} - \alpha_f$. The maximum on the left is obtained with the aid of the Schwarz inequality, discussed on p.18. One finds that the greatest value of the response at time t is:

$$\max_{u \in \mathcal{U}(\alpha_i)} x_u(t) = \frac{\alpha_i}{2m\omega^{3/2}} \sqrt{2\omega t - \sin 2\omega t} \tag{4.8}$$

So the response and failure sets are disjoint if and only if:

$$\frac{\alpha_i}{2m\omega^{3/2}} \sqrt{2\omega t - \sin 2\omega t} < \overline{f} - \alpha_f \tag{4.9}$$

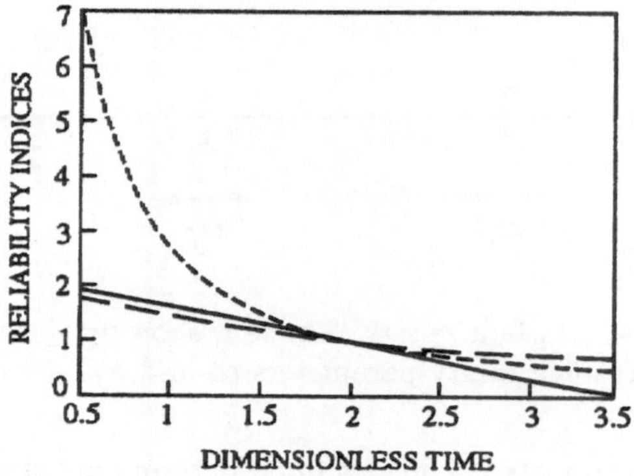

Figure 4.2: Reliability indices versus $2\omega t$. Solid: $\widehat{\alpha}(t)$; big-dash: $\widehat{\alpha}_f(t, \alpha)$; small-dash: $\widehat{\alpha}_i(t, \alpha_f)$. $\overline{f} = 2$, $\alpha_i = \alpha_f = 2m\omega^{3/2} = 1$.

We can now evaluate the reliability indices for this system.

For fixed failure uncertainty, α_f, the input reliability at time t, $\widehat{\alpha}_i(t, \alpha_f)$, is the least value of α_i for which the response and failure sets intersect. Treating (4.9) as an equality and solving for α_i yields this conditional reliability:

$$\widehat{\alpha}_i(t, \alpha_f) = \frac{(\overline{f} - \alpha_f)2m\omega^{3/2}}{\sqrt{2\omega t - \sin 2\omega t}} \tag{4.10}$$

As shown by the small-dash line in fig. 4.2, the input reliability index $\widehat{\alpha}_i(t, \alpha_f)$, for $\alpha_f = 1$, decreases monotonically in time, meaning that the amount of input uncertainty which can be tolerated diminishes in time. Conversely, for a given amount of input uncertainty α_i and failure uncertainty α_f, the system is reliable — it will operate without failure — up to a particular time t_f, which is the solution of:

$$\alpha_i = \widehat{\alpha}_i(t_f, \alpha_f) \tag{4.11}$$

Beyond instant t_f the system can reach a failure state, if the input and failure uncertainties are α_i and α_f, respectively.

We now consider $\widehat{\alpha}_f(t, \alpha_i)$. The failure reliability for fixed input uncertainty α_i is the least value of α_f for which the response and failure sets intersect. Solving for α_f in eq.(4.9) as an equality yields:

$$\widehat{\alpha}_f(t, \alpha_i) = \overline{f} - \frac{\alpha_i}{2m\omega^{3/2}}\sqrt{2\omega t - \sin 2\omega t} \tag{4.12}$$

provided the righthand side is non-negative. $\widehat{\alpha}_f(t, \alpha)$ decreases monotonically with t until no non-negative value of α_f can satisfy the condition for no-failure, as shown by the big-dash line in fig. 4.2 for $\alpha_i = 1$.

We have assumed in formulating the failure set, $\mathcal{F}(\alpha_f)$, that the failure uncertainty, α_f, is no greater than $\overline{f}$. Consequently the failure reliability index $\widehat{\alpha}_f$ cannot exceed the value $\overline{f}$. Furthermore, and also unlike the input reliability index, $\widehat{\alpha}_f(t,\alpha)$ reaches 0 at a finite time, t_0. In other words, for fixed input uncertainty α_i, the system is able to reach a failure state at any time $t > t_0$, even if the failure uncertainty is zero. This makes sense since the nominal failure state $\overline{f}$ can eventually be reached.

Now we consider the overall reliability index, which is evaluated by replacing both α_i and α_f by α in (4.9) and finding the least α for which equality holds, resulting in:

$$\widehat{\alpha}(t) = \frac{2m\omega^{3/2}\overline{f}}{2m\omega^{3/2} + \sqrt{2\omega t - \sin 2\omega t}} \tag{4.13}$$

$\widehat{\alpha}(t)$ is shown by the solid line in fig. 4.2, and displays a combination of the features of the input and failure reliability indices. We notice that all three curves coincide at $\widehat{\alpha} = \widehat{\alpha}_i = \widehat{\alpha}_f = 1$.

Examination of eqs.(4.10), (4.12) and (4.13) shows that:

$$\frac{1}{\widehat{\alpha}(t)} = \frac{1}{\widehat{\alpha}_i(t,0)} + \frac{1}{\widehat{\alpha}_f(t,0)} \tag{4.14}$$

That is, the overall reliability index, $\widehat{\alpha}(t)$, is the harmonic mean of the input and failure reliability indices at $\alpha_i = \alpha_f = 0$. From eqs.(4.10) and (4.12) we see that $\widehat{\alpha}_i(t,\alpha_f)$ and $\widehat{\alpha}_f(t,\alpha_i)$ decrease with α_f and α_i, respectively, as we should expect. This, combined with eq.(4.14), establishes the following inequality:

$$\frac{1}{\widehat{\alpha}(t)} \leq \frac{1}{\widehat{\alpha}_i(t,\alpha_f)} + \frac{1}{\widehat{\alpha}_f(t,\alpha_i)} \tag{4.15}$$

This is a special case of a more general result relating input, failure and overall reliabilities [17].

4.2 Seismic Safety of Secondary Equipment

The simple harmonic oscillator studied in the previous section is useful for representing realistic problems, as we will show now with an example based on seismic safety of buildings.

Seismically-safe design of buildings is not limited to assuring structural integrity alone. Also important is the functional integrity of critical secondary equipment such as communication units, fire-control facilities, and so on. Building codes for seismically-safe structures require that the designer guarantee specified limits to the inertial forces acting on critical secondary equipment during earthquakes [90].

In this section we will use a convex model to represent uncertainty in the temporal waveform of an earthquake excitation, and derive an expression for the maximum inertial force which can be exerted on a piece of light equipment which is dynamically coupled to a building. The convex model is chosen to include all waveforms consistent with given spectral information about the seismic ground motion. The secondary equipment is presumed to fail if the inertial force exceeds a given threshold value. We will then evaluate the reliability of this equipment with respect to the uncertainty in the seismic excitation. An analysis such as this enables the designer to choose the dynamical coupling of the equipment to the building so as to optimize the reliability of the equipment.

A major limitation of this analysis is its dependence on an accurate dynamical model for the motion of the building during an earthquake and for the dynamical coupling of the building to the equipment. This is characteristic of any model-based dynamical analysis.

An additional limitation is the paucity of information upon which the convex model is founded. However, the adverse effect of limited information about seismic variability cannot properly be viewed as a limitation of the analysis, but rather an inherent deficiency in the informational infrastructure upon which the analysis rests. To the extent that the convex model includes all earthquake ground motions consistent with existing information, the convex model is faithful to the information, without introducing additional strong assumptions about earthquake behavior. It must of course be recognized that a convex model does introduce assumptions or extrapolations beyond the raw data. A convex model contains an infinity of functions while the primary observations are surely finite.

It sometimes happens that statistical information about earthquake variability is available. This can be used in various ways to generate a hybrid robust-probabilistic reliability analysis. In chapter 8 we will discuss one form of this combined analysis.

4.2.1 Dynamics

The structure to which the equipment is attached is modelled as an N-dimensional linear elastic system with viscous damping:

$$M\ddot{x}(t) + C\dot{x}(t) + Kx(t) = Bu(t) \tag{4.16}$$

where M, C and K are constant mass, damping and stiffness matrices, respectively, $x(t)$ is the N-dimensional deflection vector, $u(t)$ is the N_I-dimensional input vector and B is a constant input matrix.

The natural frequencies of the structure are $\omega_1, \ldots, \omega_N$, and the corresponding mode-shape vectors are $v^1, \ldots, v^N$. The orthogonality and normalization properties of the mode shapes are:

$$v^{i^T} M v^j = m_i \delta_{ij}, \quad v^{i^T} K v^j = \omega_i^2 m_i \delta_{ij}, \quad v^{i^T} C v^j = 2\zeta_i \omega_i m_i \delta_{ij} \tag{4.17}$$

m_n and ζ_n are the modal mass and the damping ratio for the nth mode. We also define the damped natural frequencies as $\omega_{nd} = \omega_n \sqrt{1 - \zeta_n^2}$, and we assume that $\zeta^2 < 1$. The modal matrix is $V = [v^1, \ldots, v^N]$, so the displacement vector x is related to the vector η of modal coordinates as $x = V\eta$.

We will assume that the nth mode is dominant during seismic excitation, so the displacement of the ith node of the structure is:

$$\begin{aligned} x_i(t) &\approx v_i^n \eta_n(t) && (4.18) \\ &= \frac{v_i^n}{m_n \omega_{nd}} \int_0^t v^{n^T} Bu(\tau) \mathrm{e}^{-\zeta_n \omega_n (t-\tau)} \sin \omega_{nd}(t-\tau)\, \mathrm{d}\tau && (4.19) \end{aligned}$$

The secondary equipment is attached to the ith node of the structure. We will model this dynamical coupling as a first-order linear oscillator with mass, damping ratio and undamped natural frequency m_e, ζ_e and ω_e. Define $c_e = 2\zeta_e \omega_e m_e$, $\omega_{ed} = \omega_e \sqrt{1-\zeta_e^2}$ and $k_e = \omega_e^2 m_e$. We adopt the assumption of 'cascaded dynamics', which asserts that the motion of the equipment is driven by the floor motion, but that the dynamics of the building are uneffected by the equipment motion. The equation of motion for the equipment is:

$$m_e \ddot{y}(t) + c_e [\dot{y}(t) - \dot{x}_i(t)] + k_e [y(t) - x_i(t)] = 0 \tag{4.20}$$

where $y(t)$ is the equipment displacement with respect to the ground. For compactnes of notation we define the following quantities:

$$q(t) = \mathrm{e}^{-\zeta_n \omega_n t} [(k_e - c_e \zeta_n \omega_n) \sin \omega_{nd} t + c_e \omega_{nd} \cos \omega_{nd} t] \tag{4.21}$$

$$\sigma(t) = \mathrm{e}^{-\zeta_e \omega_e t} \sin \omega_{ed} t, \quad \gamma = \frac{\phi_i^n}{m_e m_n \omega_{ed} \omega_{nd}} \tag{4.22}$$

With these definitions, the deflection of the equipment becomes:

$$y_u(t) = \gamma \int_0^t \phi^{n^T} Bu(\theta) \underbrace{\int_\theta^t \sigma(t-\tau) q(\tau - \theta)\, \mathrm{d}\tau}_{r(t,\theta)}\, \mathrm{d}\theta \tag{4.23}$$

which defines the function $r(t, \theta)$. This function can be integrated to yield:

$$r(t,\theta) = m_e \omega_e \widetilde{r}(t,\theta) \tag{4.24}$$

where $\widetilde{r}(t,\theta)$ is a dimensionless function defined as:

$$\begin{aligned} \widetilde{r}(t,\theta) = &\frac{1}{2} \mathrm{e}^{-\zeta_n \omega_n (t-\theta)} \\ &\times [(\gamma_1 + \gamma_3) \cos \omega_{nd}(t-\theta) + (\gamma_2 - \gamma_4) \sin \omega_{nd}(t-\theta)] \\ &- \frac{1}{2} \mathrm{e}^{-\zeta_e \omega_e (t-\theta)} [-(\gamma_2 + \gamma_4) \sin \omega_{ed}(t-\theta) + (\gamma_1 + \gamma_3) \cos \omega_{ed}(t-\theta)] \end{aligned} \tag{4.25}$$

The following dimensionless coefficients are employed:

$$\gamma_1 = \frac{\beta_1\beta_3 + \beta_2\beta_{4+}}{\beta_{5+}}, \quad \gamma_2 = \frac{\beta_1\beta_{4+} - \beta_2\beta_3}{\beta_{5+}} \tag{4.26}$$

$$\gamma_3 = \frac{\beta_2\beta_{4-} - \beta_1\beta_3}{\beta_{5-}}, \quad \gamma_4 = -\frac{\beta_1\beta_{4-} + \beta_2\beta_3}{\beta_{5-}} \tag{4.27}$$

$$\beta_1 = 1 - 2\zeta_e\zeta_n\frac{\omega_n}{\omega_e}, \quad \beta_2 = 2\frac{\omega_n}{\omega_e}\zeta_e\sqrt{1-\zeta_n^2}, \quad \beta_3 = \zeta_e - \zeta_n\frac{\omega_n}{\omega_e} \tag{4.28}$$

$$\beta_{4\pm} = \sqrt{1-\zeta_e^2} \pm \frac{\omega_n}{\omega_e}\sqrt{1-\zeta_n^2}, \quad \beta_{5\pm} = \beta_3^2 + \beta_{4\pm}^2 \tag{4.29}$$

The integral $r(t,\theta)$ in eq.(4.23) always exists, but eq.(4.25) is valid only if the following relation holds:

$$(\beta_{5-})(\beta_{5+}) \neq 0 \tag{4.30}$$

We will assume throughout our calculations that this relation is valid.

In seismic applications it is usually reasonable to assume that the excitation, $u(t)$, is a scalar function, and we will do so. Define the scalar quantity:

$$\overline{\gamma} = \gamma v^{n^T} B \tag{4.31}$$

where γ is defined in eq.(4.22). Then eq.(4.23), the displacement of the equipment with respect to the ground, as a function of the ground motion becomes:

$$y_u(t) = \overline{\gamma} m_e \omega_e \int_0^t \widetilde{r}(t,\theta)u(\theta)\,d\theta \tag{4.32}$$

The inertial force acting on the secondary equipment is approximated as:

$$F_u = m_e\omega_e^2 y_u \tag{4.33}$$

We can now evaluate the maximum inertial force exerted on the secondary equipment, by finding the maximum displacement, as the seismic input varies on its convex model. This will lead to an expression for the robust reliability of the equipment.

4.2.2 Reliability with the Fourier-Envelope Model

Let us assume that $u(t) = 0$ for $t < 0$, and that $\int_0^\infty u^2(t)\,dt$ is bounded. Then the symmetrical Fourier transforms exist and are defined as:

$$u(t) = \frac{1}{\sqrt{2\pi}}\int_{-\infty}^{\infty} \overline{u}(\omega)e^{-j\omega t}\,d\omega, \quad \overline{u}(\omega) = \frac{1}{\sqrt{2\pi}}\int_0^{\infty} u(t)e^{j\omega t}\,dt \tag{4.34}$$

The Fourier-envelope convex model is the set of input functions $u(t)$ for which the norm of the Fourier transform $\overline{u}(\omega)$ is contained in a real envelope:

$$\mathcal{U}_{\text{FE}}(\alpha) = \left\{u(t) : \ |\overline{u}(\omega)|^2 \le \alpha^2 R^2(\omega)\right\} \tag{4.35}$$

where $R^2(\omega)$ is a known real envelope function, and α^2 is the uncertainty parameter representing our lack of information about the seismic input.

The real and complex parts of $\overline{u}(\omega)$ are just the Fourier cosine and sine transforms of $u(t)$, which we denote $\overline{u}_c(\omega)$ and $\overline{u}_s(\omega)$ respectively:

$$\overline{u}(\omega) = \overline{u}_c(\omega) + j\overline{u}_s(\omega) \tag{4.36}$$

$\overline{u}_c(\omega)$ and $\overline{u}_s(\omega)$ are real functions:

$$\overline{u}_c(\omega) = \frac{1}{\sqrt{2\pi}} \int_0^\infty u(t) \cos \omega t \, dt \tag{4.37}$$

$$\overline{u}_s(\omega) = \frac{1}{\sqrt{2\pi}} \int_0^\infty u(t) \sin \omega t \, dt \tag{4.38}$$

Substituting (4.36) into the first of eqs.(4.34) and recognizing that $u(t)$ is a real function one finds:

$$u(t) = \frac{1}{\sqrt{2\pi}} \int_{-\infty}^\infty [\overline{u}_c(\omega) \cos \omega t - \overline{u}_s(\omega) \sin \omega t] \, d\omega \tag{4.39}$$

Substituting this into eq.(4.32) and changing the order of integration one finds the displacement of the equipment to be:

$$y(t) = \overline{\gamma} m_e \omega_e \int_{-\infty}^\infty [\overline{u}_c(\omega)\widetilde{r}_c(t,\omega) - \overline{u}_s(\omega)\widetilde{r}_s(t,\omega)] \, d\omega \tag{4.40}$$

where we define:

$$\widetilde{r}_c(t,\omega) = \frac{1}{\sqrt{2\pi}} \int_0^t \widetilde{r}(t,\theta) \cos \omega\theta \, d\theta \tag{4.41}$$

$$\widetilde{r}_s(t,\omega) = \frac{1}{\sqrt{2\pi}} \int_0^t \widetilde{r}(t,\theta) \sin \omega\theta \, d\theta \tag{4.42}$$

To maximize $y(t)$ on $\mathcal{U}_{\text{FE}}(\alpha)$ we note that the constraint on $\overline{u}(\omega)$, at each value of ω, is:

$$\overline{u}_c^2(\omega) + \overline{u}_s^2(\omega) \le \alpha^2 R^2(\omega) \tag{4.43}$$

The Cauchy inequality is now used to find the maximum displacement as:

$$\widehat{y}(t) = \max_{u \in \mathcal{U}_{\text{FE}}} y(t) \tag{4.44}$$

$$= \alpha\overline{\gamma} m_e \omega_e \int_{-\infty}^\infty R(\omega)\sqrt{\widetilde{r}_c^2(t,\omega) + \widetilde{r}_s^2(t,\omega)} \, d\omega \tag{4.45}$$

Employing eq.(4.33), the maximum inertial force which can be exerted on the secondary equipment is:

$$\widehat{F} = m_e \omega_e^2 \widehat{y} \tag{4.46}$$

The equipment is presumed not to fail if the inertial force does not exceed the critical value F_{cr}:

$$\widehat{F} \leq F_{cr} \tag{4.47}$$

Combining the last three relations, and writing them more explicitly, we obtain the following expression for the robust reliability of the secondary equipment with respect to the input uncertainty of the seismic excitation:

$$\begin{aligned}\widehat{\alpha} &= \frac{F_{cr} m_n \omega_n \sqrt{1-\zeta_n^2}\sqrt{1-\zeta_e^2}}{m_e \omega_e^2 \phi_i^n \left(\phi^{n\,T} B\right)} \\ &\quad \times \left[\int_{-\infty}^{\infty} R(\omega)\sqrt{\widetilde{r}_c^2(t,\omega)+\widetilde{r}_s^2(t,\omega)}\,\mathrm{d}\omega\right]^{-1}\end{aligned} \tag{4.48}$$

When $\widehat{\alpha}$ is large, great input uncertainty is consistent with failure-free operation of the equipment. On the contrary, when $\widehat{\alpha}$ is small, the equipment is fragile with respect to input uncertainty and failure can occur even with low input uncertainty.

The robustness is inversely proportional to the equipment mass m_e and to the square of the natural frequency, ω_e, of the equipment-structure coupling, and proportional to the modal mass m_n and frequency ω_n of the dominant structural mode. The quantities $\widetilde{r}_c(t,\omega)$ and $\widetilde{r}_s(t,\omega)$ are functions of dimensionless coefficients depending on the the damping ratios of the equipment and of the structural mode, and on the ratio ω_n/ω_e of the structural to the equipment natural frequencies. The function $R(\omega)$ is the convex model bound on the spectrum of the input, and is based on measured spectral variability of earthquakes. One can use eq.(4.48) to assign values to the design parameters to optimize the robustness of the equipment. Similarly, this relation enables one to identify those physical parameters which are dominant in controlling the reliability of the equipment.

4.3 Multi-Dimensional Vibrating Structures

We have considered a simple harmonic oscillator in general terms in section 4.1, and in application to seismic reliability in section 4.2. We now consider the reliability of a linear elastic structure of arbitrary dimension, subject to uncertain inputs and uncertain failure states.

4.3.1 Formulation

Consider an L-dimensional linear vibrating system represented by:

$$M\ddot{y}(t) + Ky(t) = Bu(t) \tag{4.49}$$

where M is the mass matrix and K is the stiffness matrix, $y(t)$ is the vector of displacements of the nodes of the structure, $u(t)$ is an uncertain band-limited scalar input function and $B \in \Re^{L\times 1}$ is the input matrix.

For $t \in [0, T]$, let the band-limited inputs be represented as:

$$u(t) = \sum_{n=n_1}^{n_2} \gamma_n \sin n\pi t/T = \gamma^T \sigma(t) \tag{4.50}$$

Define $N = n_2 - n_1 + 1$, so γ is the N-vector of Fourier coefficients and $\sigma(t)$ is an N-vector of sine functions. Let the uncertainty of the input Fourier coefficients be represented by an ellipsoid-bound convex model centered at the origin:

$$\mathcal{U}(\alpha_i) = \{\gamma : \ \gamma^T Q \gamma \le \alpha_i^2\} \tag{4.51}$$

where Q is a real, symmetric, positive definite $N \times N$ matrix.

The natural frequencies of the structure are $\omega_1, \ldots, \omega_L$ and the corresponding mode-shape vectors are $v^1, \ldots, v^L$. The modal matrix is $V = [v^1, \ldots, v^L]$. The orthogonality and normalization properties of the mode shapes are expressed in eq.(4.17).

The mode shapes are linearly independent, so the displacement vector $y(t)$ can be decomposed into modal components as:

$$y(t) = V\eta(t) \tag{4.52}$$

The modal amplitudes $\eta_n(t)$ are dynamically uncoupled and, for zero initial conditions $\eta_n(0) = \dot{\eta}_n(0) = 0$, one finds:

$$\eta_n(t) = \frac{1}{m_n\omega_n} \int_0^t v^{n^T} Bu(\tau) \sin\omega_n(t-\tau)\,\mathrm{d}\tau \tag{4.53}$$

For convenience let us define the functions:

$$z_n(t) = \frac{1}{m_n\omega_n} \sin\omega_n t \tag{4.54}$$

which are the elements of a diagonal matrix $Z(t) = \mathrm{diag}[z_1(t), \ldots, z_N(t)]$. Combining eqs.(4.53) and (4.54), the vector of modal coordinates can be written as an explicit function of the input and the modal properties as follows:

$$\eta(t) = \int_0^t Z(t-\tau) V^T Bu(\tau)\,\mathrm{d}\tau \tag{4.55}$$

Expressing the input as $u(\tau) = \sigma^T(\tau)\gamma$; this becomes:

$$\eta_\gamma(t) = \underbrace{\left(\int_0^t Z(t-\tau)V^T B\sigma^T(\tau)\,\mathrm{d}\tau\right)}_{\zeta^T(t)} \gamma \tag{4.56}$$

$$= \zeta^T(t)\gamma \tag{4.57}$$

which defines the $L \times N$ matrix $\zeta^T(t)$. The subscript on η_γ expresses the dependence of the modal response upon the uncertain input Fourier coefficients, γ.

Each input function $u(t) = \sigma^T(t)\gamma$ generates a response in modal-coordinate space, $\eta_\gamma(t)$, and the set of all these modal output vectors is the *modal response set,* which can be represented:

$$\mathcal{X}(\alpha_i) = \left\{\eta_\gamma(t) = \zeta^T(t)\gamma, \quad \text{for all} \quad \gamma \in \mathcal{U}(\alpha_i)\right\} \tag{4.58}$$

As in section 4.1, we identify output states which constitute failure if reached by the system. Let $f(t)$ represent a displacement vector which, if reached by the system, constitutes failure. That is, the system is considered to have failed at time t if $y(t) = f(t)$. We will represent output failure states in Fourier expansion as:

$$f(t) = \sum_{n=n_1}^{n_2} \psi_n \sin n\pi t/T \tag{4.59}$$

Collect the coefficient vectors ψ_n into a matrix $\Psi = [\psi_{n_1}, \ldots, \psi_{n_2}] \in \Re^{L\times N}$, so the failure state can be expressed as:

$$f(t) = \Psi\sigma(t) \tag{4.60}$$

The Fourier coefficients of the nominal failure state are $\overline{\Psi}$. However, there is uncertainty in the failure state which is expressed in terms of uncertainty in its Fourier coefficients. Using a euclidean-norm bound convex model, the *failure set* is:

$$\mathcal{F}(\alpha_f) = \left\{f(t) = \Psi\sigma(t): \; \left\|\Psi - \overline{\Psi}\right\|^2 \le \alpha_f^2\right\} \tag{4.61}$$

where the euclidean norm for matrices is:

$$\|\Psi - \overline{\Psi}\|^2 = \sum_{n=n_1}^{n_2} \left(\psi_n - \overline{\psi}_n\right)^T \left(\psi_n - \overline{\psi}_n\right) \tag{4.62}$$

4.3.2 Reliability: Hyperplane Separation

For input uncertainty α_i and failure uncertainty α_f, the condition for no-failure is that the response and failure sets are disjoint:

$$\mathcal{X}(\alpha_i) \cap \mathcal{F}(\alpha_f) = \emptyset \tag{4.63}$$

In other words, the *input reliability* $\widehat{\alpha}_i(t, \alpha_f)$, for fixed failure uncertainty α_f, is the upper bound of values of α_i for which this disjointness holds. Similarly, the *failure reliability* $\widehat{\alpha}_f(t, \alpha_i)$, for fixed input uncertainty α_i, is the upper bound of α_f values at which the response and failure sets are disjoint. The *overall reliability* is obtained by replacing both α_i and α_f by α in (4.63), and seeking the value of α at which the response and failure sets are just tangent. This is precisely the multi-dimensional analog of eq.(4.6).

To determine the robust reliability indices we must establish conditions for disjointness of the convex sets $\mathcal{X}$ and $\mathcal{F}$. This is readily done using the hyperplane separation theorem discussed in chapter 2. These sets are disjoint if and only if there is a hyperplane which separates the response set $\mathcal{X}$ from the failure set $\mathcal{F}$. Since the response set is centered at the origin, the hyperplane separation theorem implies that (4.63) holds if and only if there is a vector $w \in \Re^L$ such that:

$$\max_{\eta \in \mathcal{X}(\alpha_i)} w^T \eta < \min_{f \in \mathcal{F}(\alpha_f)} w^T f \tag{4.64}$$

This can be written explicitly in terms of the uncertain Fourier coefficients γ and Ψ. Recall that elements of the response set can be written $\eta = \zeta^T \gamma$, while elements of the failure set are $f = \Psi\sigma$. Thus (4.64) becomes:

$$\max_{\gamma \in \mathcal{U}(\alpha_i)} w^T \zeta^T(t)\gamma < \min_{\Psi \in \mathcal{F}(\alpha_f)} w^T \Psi\sigma(t) \tag{4.65}$$

These extrema are readily found using Lagrange optimization, and (4.64) can be expressed:

$$\alpha_i \frac{\sqrt{w^T \zeta^T Q^{-1} \zeta w}}{||w||} < \frac{w^T \overline{\Psi}\sigma}{||w||} - \alpha_f ||\sigma|| \tag{4.66}$$

where $||\cdot||$ is the euclidean norm for vectors. The response and failure sets are disjoint if and only if there is a vector w for which this inequality holds.

4.3.3 Input Reliability

Let us consider the application of relation (4.66) for determining the input reliability $\widehat{\alpha}_i(t, \alpha_f)$ when there is no failure uncertainty, so $\alpha_f = 0$. The input reliability is the greatest value of α_i for which there is a real L-vector w satisfying (4.66) as an equality:

$$\alpha_i \frac{\sqrt{w^T \zeta^T Q^{-1} \zeta w}}{||w||} = \frac{w^T \overline{\Psi}\sigma}{||w||} \tag{4.67}$$

Let us suppose that $\zeta^T Q^{-1} \zeta$ is a positive definite matrix, which will usually be the case. Denote its eigenvalues and orthonormal eigenvectors as $\lambda_1, \ldots, \lambda_L$ and $\theta_1, \ldots, \theta_L$ respectively. We can project w and $\overline{\Psi}\sigma$ on the eigenvectors as:

$$w = \sum_{i=1}^{L} \nu_i \theta_i \quad \text{and} \quad \overline{\Psi}\sigma = \sum_{i=1}^{L} \mu_i \theta_i \tag{4.68}$$

Denote $\nu^T = (\nu_1, \ldots, \nu_L)$, $\mu^T = (\mu_1, \ldots, \mu_L)$ and $\Lambda = \text{diag}(\lambda_1, \ldots, \lambda_L)$. Thus:

$$w^T \zeta^T Q^{-1} \zeta w = \nu^T \Lambda \nu \tag{4.69}$$

and

$$w^T \overline{\Psi}\sigma = \nu^T \mu \tag{4.70}$$

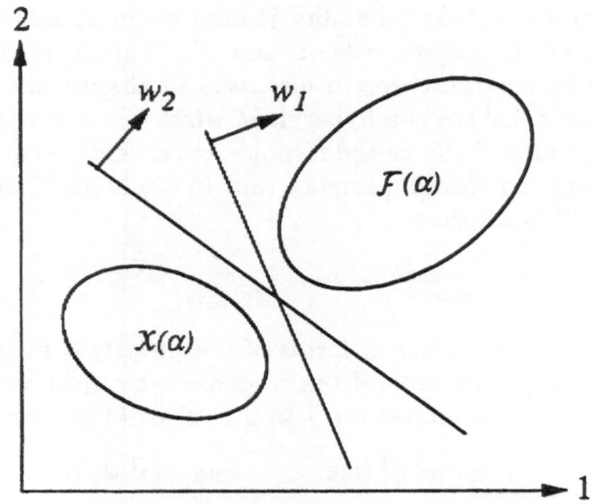

Figure 4.3: Schematic representation of disjoint response and failure sets.

The uncertainty parameter α_i must be non-negative, so, squaring (4.67) we see that the input reliability is the greatest value of α_i for which there is a unit vector ν such that:

$$\alpha_i^2 = \frac{\nu^T \mu\mu^T \nu}{\nu^T \Lambda \nu} \tag{4.71}$$

$$= \frac{\xi^T \Lambda^{-1/2} \mu\mu^T \Lambda^{-1/2} \xi}{\xi^T \xi} \tag{4.72}$$

where we have defined $\xi = \Lambda^{1/2}\nu$. The maximum of the Raleigh quotient in eq.(4.72) is the greatest eigenvalue of the matrix in the numerator, which is $\mu^T \Lambda^{-1} \mu$. Thus the input reliability, for zero failure uncertainty, is:

$$\widehat{\alpha}_i(t,0) = \max \operatorname{eig} \left[\Lambda^{-1/2} \mu\mu^T \Lambda^{-1/2}\right] = \mu^T \Lambda^{-1} \mu \tag{4.73}$$

This equation establishes the relation between the dynamics of the system as embodied in ζ, the frequencies of excitation represented by the functions σ, the structure of the input uncertainty set as determined by Q, and the nominal failure state $\overline{\Psi}$.

4.4 Modal Reliability

In subsection 4.3.2 we used the hyperplane-separation theorem to determine the robust reliability of a multi-dimensional vibrational system. The disjointness relation, (4.63), is the condition for no-failure, and it holds if and only

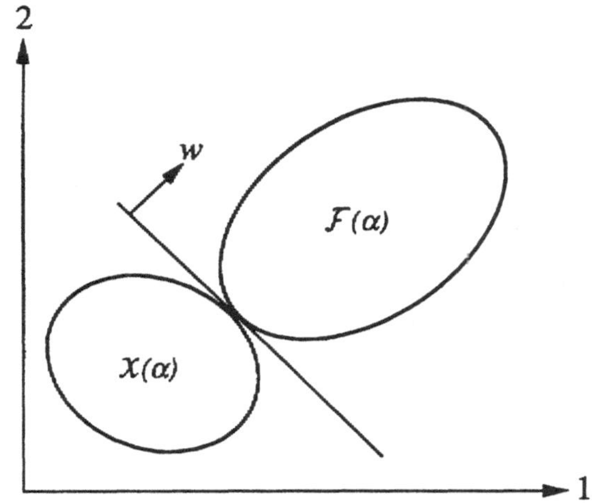

Figure 4.4: Schematic representation of tangent response and failure sets.

if there is a vector w satisfying inequality (4.64). In this section we will develop an interpretation of the vector w, which will lead to the idea of relative reliability of dynamic modes.

4.4.1 Formulation

The uncertain response set $\mathcal{X}(\alpha)$ is disjoint from the uncertain failure set $\mathcal{F}(\alpha)$ if and only if there is a vector w such that:

$$\max_{x \in \mathcal{X}(\alpha)} w^T x < \min_{f \in \mathcal{F}(\alpha)} w^T f \tag{4.74}$$

The overall reliability, $\widehat{\alpha}$, is the least upper bound of α-values for which these sets are disjoint. For any smaller value of α these sets are disjoint, and a range of hyperplanes will separate them, as shown in fig. 4.3. When $\alpha = \widehat{\alpha}$ the sets just touch. Let w be a vector defining a hyperplane tangent to the point of intersection, as in fig. 4.4. This vector will be unique if the boundary of the sets is smooth at the point of intersection, which is the usual case.

Now, what information does this tangent vector w contain? Consider the schematic situation in fig. 4.5, where w is parallel to one axis and perpendicular to the other: $w^T = (1, 0)$. The sets have been expanded until they touch, but it is evident that expansion along axis 2 (if we could separate it from expansion along axis 1) would have no effect on the reliability, which is controlled entirely by expansion along axis 1. Let us say then that the coordinate of axis 2 is "much more reliable" than the coordinate of axis 1. If, for instance, these are modal coordinates, we might say that the second mode is much more reliable than the first, since unlimited uncertainty is tolerable "along the second mode" before failure occurs, while uncertainty "along the

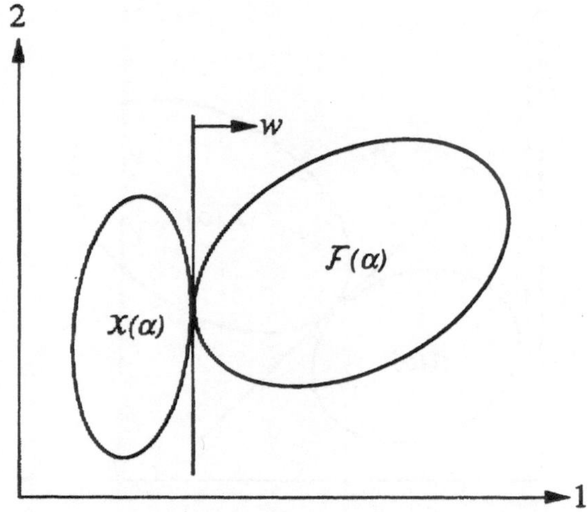

Figure 4.5: Schematic representation of tangent response and failure sets with a perpendicular separating hyperplane: $w^T = (1, 0)$.

first mode" controls the reliability index $\widehat{\alpha}$. This is clearly a qualitative statement, since the sets expand in all directions at once, but it does represent the fact that the relative dispositions and shapes of the sets cause one mode to dominate over the other in determining the reliability.

Returning to fig. 4.4, we notice that the tangent hyperplane is inclined at 45° to the axes: $w^T = (1, 1)$. This indicates, in contrast to the situation in fig. 4.5, that the axes contribute equally to the reliability index.

In general, we can say that the elements of a vector w, which defines the tangent hyperplane separating response and failure sets, are inversely related to the relative reliability of the degrees of freedom of the corresponding axes.

4.4.2 Coordinate Transformations

An immediate consequence of the hyperplane separation theorem is that the reliability indices are unchanged by an orthogonal transformation of the response space. Let Γ be an orthogonal transformation of the response space, transforming the response vector from y into x, according to the relation $x = \Gamma y$, and transforming the failure states from h to f as $f = \Gamma h$. For example, we might choose Γ as the modal matrix of a linear elastic system, in which case y and h are the modal coordinates of the response x and of the failure state f, respectively.

If $\mathcal{X}$ is the response set in the x coordinate system, let $\mathcal{Y}$ denote the response set in the transformed space of y vectors:

$$\mathcal{Y} = \Gamma^T \mathcal{X} = \{y : \ y = \Gamma^T x, \text{ for all } x \in \mathcal{X}\} \tag{4.75}$$

We can define the transformation of the failure set $\mathcal{F}$ similarly as $\mathcal{H} = \Gamma^T \mathcal{F}$. The hyperplane-separation condition for disjointness of $\mathcal{X}$ from $\mathcal{F}$ is the existence of a vector w such that relation (4.74) holds. Defining $v = \Gamma^T w$ shows that relation (4.74) is equivalent to:

$$\max_{y \in \mathcal{Y}} v^T y \ < \ \min_{h \in \mathcal{H}} v^T h \tag{4.76}$$

since $v^T y = w^T \Gamma y = w^T x$ and $v^T h = w^T \Gamma h = w^T f$. In other words, $\mathcal{X}(\alpha)$ and $\mathcal{F}(\alpha)$ are disjoint if and only if $\mathcal{Y}(\alpha)$ and $\mathcal{H}(\alpha)$ are disjoint. Consequently, the reliability — the value of α at which the sets are tangent — is unchanged by the coordinate transformation.

However, the relative reliabilities of the degrees of freedom in the two coordinate systems may be quite different. In light of the equivalence of relations (4.74) and (4.76), we can derive the relative reliabilities of one set of coordinates from the relative reliabilities of another. For example, suppose that w in (4.74) defines the tangent separating hyperplane in the response space x whose coordinates are nodal displacement variables. The elements of w express the relative reliabilities of these displacement variables. Let Γ be the modal matrix, so that $y = \Gamma^T x$ is the modal coordinate vector. Then the tangent separating hyperplane in modal coordinate space is $v = \Gamma^T w$. The elements of v express the relative reliabilities of the modal degrees of freedom.

Furthermore, we can always define a coordinate transformation for which the reliability is "controlled" by a single variable. We do this as follows. Let w be a unit vector defined as in the previous paragraph: normal to the tangent separating hyperplane in nodal response space. Let $w_2, \ldots, w_N$ be an orthonormal basis for the null space of w. That is, $w^T w_n = 0$ for $n = 2, \ldots, N$, and $w_n^T w_m = \delta_{nm}$. Now define a matrix $H = [w, w_2, \ldots, w_N]$, which is an orthogonal matrix. That is, $H^T H = I$. Let e^1 be a vector having a '1' in the first element and '0' elsewhere. We see that H^T defines a linear map which transforms w into e^1, since $\mathrm{e}^1 = H^T w$. Now define a transformed response vector by $q = H^T x$. Referring to the equivalence of relations (4.74) and (4.76) we see that the tangent separating hyperplane for coordinate system q is e^1. Thus the reliability is controlled entirely by the first coordinate of q, which can be expressed as the following linear combination of the nodal displacements: $q_1 = \mathrm{e}^{1^T} H^T x$.

4.5 Axially Loaded Thin-Walled Shell With Imperfect Initial Shape

In chapter 3 we considered the reliability of cylindrical shells with radial load and azimuthally varying geometrical imperfections. In this section we consider a complementary problem, in treating the axially-loaded shell with rotationally symmetric imperfections. In both cases, the imperfection-sensitivity of the thin-walled structure motivates the reliability analysis. Previously we employed a maximum-deflection criterion for failure, while now we will apply global energetic considerations to define failure.

4.5.1 Dynamics

The differential equation for rotationally symmetric motion of a thin-walled cylindrical shell is:

$$D\frac{\partial^4 w}{\partial x^4} + P\frac{\partial^2 w}{\partial x^2} + \rho h\frac{\partial^2 w}{\partial t^2} + \frac{Eh}{R^2}w = -P\frac{\partial^2 w_0}{\partial x^2} \tag{4.77}$$

where x is the axial coordinate, t is the time, $w_0(x)$ is the initial imperfection of the shell at height x, $w(x,t)$ is the additional shell displacement, $D = Eh^3/12(1-\nu^2)$ is the flexural stiffness, E is Young's modulus, h is the shell-wall thickness, R is the shell radius, ν is Poisson's ratio, P is the axial load and ρ is the density of the shell material.

We will employ the following non-dimensional quantities:

$$\xi = x/L, \quad \tau = \omega_1 t, \quad \lambda = P/P_{\mathrm{cl}} \tag{4.78}$$

where L is the shell length, ω_1 is the first natural frequency of the shell and P_{cl} is the lowest classical buckling load of the ideal shell:

$$\omega_1^2 = \frac{D\pi^4}{\rho h L^4} + \frac{E}{\rho R^2}, \quad P_{\mathrm{cl}} = \frac{Eh^2}{R\sqrt{3(1-\nu^2)}} \tag{4.79}$$

In addition, we define non-dimensional additional displacement and initial imperfection:

$$u = w/h, \quad u_0 = w_0/h \tag{4.80}$$

The initial imperfection function can be expanded in a Fourier sine series:

$$u_0(\xi) = \sum_{n=1}^{\infty} a_n \sin n\pi\xi \tag{4.81}$$

The differential equation of motion of the shell, for simply supported ends, can be solved to express the total normalized displacement, $v(\xi,\tau)$, in terms of the Fourier coefficients of the initial imperfection, as follows [37]:

$$v(\xi,\tau) = u_0(\xi) + u(\xi,\tau) = \sum_{n=1}^{\infty} a_n[1+\psi_n(\tau)]\sin n\pi\xi \tag{4.82}$$

where the functions $\psi_n(\tau)$ are defined as:

$$\psi_n(\tau) = \begin{cases} \dfrac{\lambda\overline{\gamma}a_n}{p_n}[\cosh(r_n\tau)-1], & p_n > 0 \\ \dfrac{\lambda\overline{\gamma}a_n}{p_n}[\cos(r_n\tau)-1], & p_n < 0 \\ \dfrac{\lambda\overline{\gamma}n^2\pi^2\tau^2 a_n}{2(\beta+\pi^2)}, & p_n = 0 \end{cases} \tag{4.83}$$

where:

$$\overline{\gamma} = \frac{4L^2}{Rh}\sqrt{3(1-\nu^2)}, \quad \beta = \frac{EhL^4}{DR^2}, \tag{4.84}$$

$$p_n = \lambda\overline{\gamma} - n^2\pi^2 - \frac{\beta}{n^2\pi^2}, \quad r_n = \frac{n\pi}{\sqrt{\beta+\pi^4}}\sqrt{|p_n|} \tag{4.85}$$

4.5.2 Fourier Ellipsoid Bound

Let us suppose that only N spatial modes appear in the initial imperfections, with indices $m_1, \ldots, m_N$. Thus the total normalized displacement, $v(\xi,\tau)$ in eq.(4.82), is the sum of N terms, and can be expressed as the inner product of the vector $a^T = (a_{m_1}, \ldots, a_{m_N})$ of uncertain Fourier coefficients with a vector $\sigma(\xi,\tau)$ of known functions:

$$v(\xi,\tau) = a^T\sigma(\xi,\tau)[1+\psi_n(\tau)]\sin n\pi\xi \tag{4.86}$$

where:

$$\sigma_n(\xi,\tau) = [1+\psi_n(\tau)]\sin n\pi\xi \tag{4.87}$$

The uncertainty in the Fourier coefficients of the initial imperfections are represented by an ellipsoid-bound convex model:

$$\mathcal{A}(\alpha) = \{a: \ a^T W a \le \alpha\} \tag{4.88}$$

The strain energy in bending of a thin shell is a quadratic function of the curvature of the surface of the shell, while the strain energy of stretching of the middle surface of the shell is quadratic in the strains [98]. In light of the linear relation between the displacement v and the Fourier coefficients a, these quadratic functions are also quadratic in a.

As an example, consider the integral over the shell surface of the squared curvature in the axial direction. Using dots to represent differentiation with respect to ξ and employing eq.(4.86) one finds:

$$\int_0^{2\pi}\int_0^1 \left(\frac{\partial^2 v}{\partial \xi^2}\right)^2 d\xi\, d\theta = 2\pi\int_0^1 a^T\ddot{\sigma}(\xi,\tau)\ddot{\sigma}^T(\xi,\tau)a\, d\xi \tag{4.89}$$

$$= a^T \underbrace{\left[2\pi\int_0^1 \ddot{\sigma}(\xi,\tau)\ddot{\sigma}^T(\xi,\tau)\, d\xi\right]}_{S} a \tag{4.90}$$

$$= a^T S a \tag{4.91}$$

where S is a known real symmetric positive definite matrix. Thus the surface integral of the square of the axial curvature is quadratic in the Fourier coefficients.

We will study the reliability of the shell with a quadratic failure function based on eq.(4.91). That is, failure will be assumed to occur when a quadratic function of the imperfection coefficients exceeds a critical value:

$$a^T Ra > E_{\rm cr} \tag{4.92}$$

where R is real, symmetric and positive definite. R will be the matrix S of eq.(4.90) if we consider only bending strain, or it will be a more complicated matrix if we consider both bending and stretching.

To determine the reliability we must establish the maximum of $a^T Ra$ subject to the quadratic constraint inherent in the convex model:

$$\max_a a^T Ra \quad \text{subject to} \quad a^T Wa \le \alpha \tag{4.93}$$

First of all, it is evident that the maximum occurs with an a-vector for which $a^T Wa = \alpha$, namely, a vector on the boundary of $\mathcal{A}(\alpha)$. Any a-vector in the interior of the ellipsoid would not give a maximum of $a^T Ra$, since $a^T Ra$ would be increased by lengthening that interior vector out to the boundary. So, we may replace the inequality in (4.93) by an equality.

Now the method of undetermined Lagrange multipliers yields a solution. Adjoin the constraint, $a^T Wa = \alpha$, to the function we seek to maximize:

$$J = a^T Ra + \mu[\alpha - a^T Wa] \tag{4.94}$$

Differentiation with respect to a yields a necessary condition for an extremum:

$$0 = \frac{dJ}{da} = 2Ra - 2\mu Wa \tag{4.95}$$

Exploiting the fact that W is positive definite, this can be re-arranged as:

$$0 = W^{1/2}\left[W^{-1/2}RW^{-1/2} - \mu I\right] W^{1/2}a \tag{4.96}$$

$W^{1/2}a$ must be an eigenvector of the real symmetric matrix $W^{-1/2}RW^{-1/2}$, and μ must be the corresponding eigenvalue. Employing the constraint to determine the normalization of the eigenvectors, we find that the maximum we seek is the greatest eigenvalue of the matrix $W^{-1/2}RW^{-1/2}$:

$$\max_{a\in\mathcal{A}(\alpha)} a^T Ra = \alpha \max \text{ eig } \left[W^{-1/2}RW^{-1/2}\right] \tag{4.97}$$

Combining this with the failure criterion, (4.92), yields the following expression for the robust reliability:

$$\widehat{\alpha} = \frac{E_{\rm cr}}{\max \text{ eig } \left[W^{-1/2}RW^{-1/2}\right]} \tag{4.98}$$

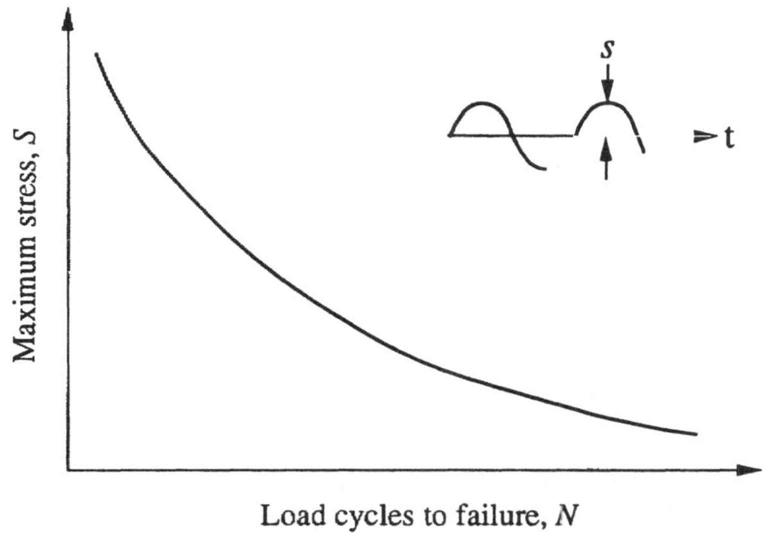

Figure 4.6: Schematic S-N curve for harmonic loading.

Example 1 Let the convex model of eq.(4.88) be a sphere, so W is the identity matrix. For the quadratic criterion $a^T Ra$ we will use the strain energy in axial bending, so instead of R we use S in eq.(4.90), which is a diagonal matrix whose nth diagonal element is

$$S_{nn} = 2\pi \int_0^1 \left[\ddot{\sigma}_n(\xi)\right]^2 \mathrm{d}\xi \tag{4.99}$$

$$= n^4\pi^5[1+\psi_n(\tau)]^2 \tag{4.100}$$

The robustness, eq.(4.98), becomes:

$$\widehat{\alpha} = \frac{E_{\mathrm{cr}}}{\max_n \; n^4\pi^5\left[1+\psi_n(\tau)\right]^2} \tag{4.101}$$

The quantity ψ_n expresses the buckling dynamics of the nth mode of the shell, so this relation shows that the robust reliability of the shell is controlled by a critical or weakest mode. ■

4.6 Fatigue Failure and Reliability With Uncertain Loading

The mechanical properties of a solid structure degrade slowly in time when it is subjected to cyclical loads at stress levels well below the yield stress of the material. This degradation process is called *high-cycle fatigue*, and can lead

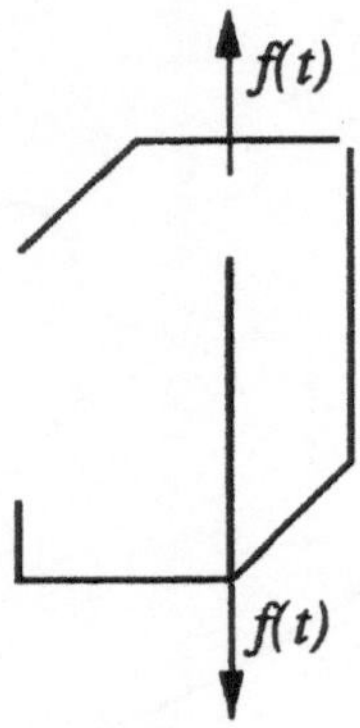

Figure 4.7: A solid element subject to time-varying load.

ultimately to failure of the material by the evolution of many small cracks or the eventual catastrophic growth of a single large crack. The classical laboratory assessment of damage evolution by fatigue is represented by the S–N curve, as in fig. 4.6, which expresses the amplitude, S, of the cyclic stress versus the number, N, of load cycles to failure. Such curves are measured in idealized laboratory conditions, where the load cycles are carefully controlled. In practice of course the load history of a real engineering system is uncertain and quite variable, unlike the perfect harmonic loading used in measuring an S–N curve. A bridge, for instance, experiences complex repeated loads due to traffic as well as perhaps wind or waves. The cutting tool of a milling machine undergoes complicated repetitive loads resulting from the dynamic vibrational response induced by its stressful contact with the work piece. Other examples of complex quasi-periodic or repetitive but uncertain load patterns abound in many mechanical applications.

In this section we will develop a phenomenological model for damage evolution and failure by high-cycle fatigue, and we will represent uncertainty in the load history with a convex model. This will serve as the basis for our analysis of the reliability with respect to fatigue failure.

4.6.1 Damage Evolution

Consider a small element in a machine or solid structure, subjected to time-varying axial load $f(t)$, as shown in fig. 4.7. One common model for approximately representing small-amplitude vibration of this element is the viscously-damped mass-spring system, whose differential equation is:

$$m\ddot{x}(t) + c\dot{x}(t) + kx(t) = f(t) \tag{4.102}$$

where m, c and k are constant positive mass, damping and stiffness coefficients respectively, $x(t)$ is the deflection and $f(t)$ is the load. The natural frequency is $\omega = \sqrt{k/m}$ and the damping ratio is $\zeta = c/2\omega$, which we assume is less than unity. The damped natural frequency is $\omega_d = \omega\sqrt{1-\zeta^2}$.

For zero initial conditions and sub-critical damping $\zeta^2 < 1$, one can use Duhamel's convolution theorem to express the displacement as:

$$x(t) = \frac{1}{m\omega_d}\int_0^t f(\tau)\mathrm{e}^{-\zeta\omega(t-\tau)} \sin\omega_d(t-\tau)\,\mathrm{d}\tau \tag{4.103}$$

The damping and stiffness parameters, c and k, change slowly as damage accumulates in the fatigue process which results from repeated loading. A *duty cycle* is a short duration T during which the system undergoes many vibrations but in which a very small increment of damage accumulates. We will assume that c and k are constant over durations T, and we will evaluate their long-range damage evolution by evaluating the small damage increments and revising the model parameters at the end of each duty cycle. We are not assuming that the system evolution is described by linear dynamic equations. We only assume that the oscillation-dynamics are linear during each short duration T.

Numerous micromechanical models have been developed for understanding the evolution of load-related damage in structures. In particular, the evolution of fatigue damage has been fruitfully associated with vibro-acoustical dissipation of energy [15, 27, 28, 72]. We apply this idea to the vibration model of eq.(4.102), in which energy is introduced by the load, flows through the inertial degree of freedom, and is dissipated. Some of this dissipated energy results in damage, leading to fatigue.

The dissipative force acting on the system at any instant is $c\dot{x}$ [N]. An infinitesimal displacement $\mathrm{d}x$ does $c\dot{x}\mathrm{d}x$ [J] of work on the system. This displacement occurs in a time interval $\mathrm{d}t$, so the rate of work on the system is $c\dot{x}\mathrm{d}x/\mathrm{d}t = c\dot{x}^2$ [W]. The accumulated energy loss in a short duty cycle of duration T during which the physical parameters $p = (c, k)$ of the system remain unchanged, is:

$$E(p) = \int_0^T c\dot{x}^2(t)\,\mathrm{d}t \tag{4.104}$$

The associated damage is modelled as a geometric function of the dissipated energy:

$$\delta(p) = \gamma\,[E(p)]^\nu \tag{4.105}$$

where γ and ν are positive constants.

We need expressions for variation of the model parameters c and k with the total accumulated damage. The dominant assumption which we will make is that, while the accumulated normalized damage, Δ, depends on the load history, $f(t)$, the model parameters c and k depend only on the magnitude of the accumulated damage. With this and some additional assumptions [15] one can show that the damage evolution must be exponential:

$$c(\Delta) = \mathrm{e}^{-\beta_c\Delta}c, \quad k(\Delta) = \mathrm{e}^{-\beta_k\Delta}k \tag{4.106}$$

The parameters β_c and β_k are scalars which need to be determined either from empirical data or from fundamental physical considerations. Failure occurs when Δ reaches the critical value, Δ_{cr}.

4.6.2 Uncertain Load Histories and Maximum Damage Increment

The input which drives the system and generates the dynamical fatigue aging is subject to uncertainty. We will model this uncertainty with the Fourier ellipsoid-bound convex model. Assume that the input can be represented during a duty cycle $[0, T]$ by a sum of harmonic functions:

$$f(t) = \sum_{n=1}^{n_f} \xi_n \cos \frac{k_n \pi t}{T} \tag{4.107}$$

where ξ_n is the Fourier coefficient of the nth temporal mode of the input and k_n is an integer. Concatenate the Fourier coefficients in a vector: $\xi^T = (\xi_1, \ldots, \xi_{n_f})$. The Fourier ellipsoid-bound convex model is the following set of allowed values of the Fourier coefficient vector:

$$\mathcal{F}_{\mathrm{feb}} = \{\xi : \ (\xi - \overline{\xi})^T W (\xi - \overline{\xi}) \leq \alpha^2\} \tag{4.108}$$

W is a real symmetric positive definite matrix which determines the shape of the ellipsoid within which the Fourier-coefficient vectors vary. The uncertainty parameter α^2 determines the size of the ellipsoid. $\overline{\xi}$ is the center point of the ellipsoid and is based on the nominal or typical input; it may be zero.

For later use we define the quantities Z_{mn} for $m, n = 1, \ldots, n_f$:

$$Z_{mn} = c \int_0^T \left[\int_0^t \dot{G}(t-\tau) \cos \frac{k_m \pi \tau}{T} \, \mathrm{d}\tau \right] \left[\int_0^t \dot{G}(t-\tau) \cos \frac{k_n \pi \tau}{T} \, \mathrm{d}\tau \right] \mathrm{d}t \tag{4.109}$$

where:

$$G(t-\tau) = \frac{1}{m\omega_d} \mathrm{e}^{-\zeta\omega(t-\tau)} \sin \omega_d (t-\tau) \tag{4.110}$$

and $\dot{G}(t-\tau) = \mathrm{d}G(t-\tau)/\mathrm{d}t$. Define Z as the n_f-dimensional square matrix constructed from the terms Z_{mn}, m, $n = 1, \ldots, n_f$. Note that Z is real and symmetric and depends on m, c and k.

We can now evaluate the maximum increment of damage which can accrue during a duty cycle, for any load-history allowed by the convex model.

Substituting eq.(4.107) into eq.(4.103) and differentiating, one finds that the energy loss in a duty cycle, eq.(4.104), becomes:

$$E(p) = \xi^T Z \xi \tag{4.111}$$

where the elements of the matrix Z are defined in eq.(4.109). Thus the increment of energy loss E during a duty cycle is quadratic in the Fourier coefficients of the uncertain load history. The Fourier coefficients of the load are constrained by the convex model, so, to find the maximum increment of damage in a duty cycle, we must seek the maximum of $E(p)$ on the set $\mathcal{F}_{\text{feb}}$:

$$\widehat{E}(p) = \max_{\xi \in \mathcal{F}_{\text{feb}}} \xi^T Z \xi \tag{4.112}$$

Combining this with eq.(4.105) we have the greatest damage which can occur in a duration T:

$$\widehat{\delta}(p) = \gamma \left[\widehat{E}(p)\right]^{\nu} \tag{4.113}$$

4.6.3 The Least-Lifetime Recursion

Starting from a given amount of damage, Δ, our aim is to evaluate $N(\Delta)$: the least number of duty cycles which results in failure, driven by a convex model of load histories. From this we will be able to evaluate the fatigue reliability. Failure can be variously defined, not necessarily as ultimate catastrophe. For example, one may choose the transition between different regimes of damage evolution, such as the transition from micro-crack growth to growth of a single dominant crack. The amount of damage is defined implicitly with respect to a reference model. Our analysis is based on the following assumptions:

1. Failure occurs when the damage reaches a critical value, Δ_{cr}.

2. The damage increases continuously from Δ to Δ_{cr}. That is, the increments of damage accumulating in each duty cycle are very small compared to the critical value, so that variation of the accumulated damage with age can be treated as a continuously increasing process.

3. Starting from a system whose accumulated damage is Δ, the least time to failure, $N(\Delta)$, depends only on Δ.

These assumptions imply that $N(\Delta)$ can be calculated from a recursive relation [15] which we explain as follows. Suppose the number of duty cycles to failure, starting from damage level $\Delta_{\text{cr}} - y$, is n:

$$N(\Delta_{\text{cr}} - y) = n \tag{4.114}$$

Then failure occurs after $n+1$ cycles, starting from a slightly lower damage level $\Delta_{\text{cr}} - y - \varepsilon$, if the maximum damage increment in one cycle is sufficient to raise the level of damage to $\Delta_{\text{cr}} - y$. That is:

$$N(\Delta_{\text{cr}} - y - \varepsilon) = n + 1 \quad \text{if} \quad \Delta_{\text{cr}} - y - \varepsilon + \widehat{\delta}(\Delta_{\text{cr}} - y - \varepsilon) = \Delta_{\text{cr}} - y \tag{4.115}$$

Simplifying the expression on the right, this becomes:

$$N(\Delta_{\text{cr}} - y - \varepsilon) = n + 1 \quad \text{if} \quad \varepsilon = \widehat{\delta}(\Delta_{\text{cr}} - y - \varepsilon) \tag{4.116}$$

This, together with eq.(4.114), constitute recursive relations for the least number of duty cycles to failure. Defining $\Delta = \Delta_{cr} - y$, these recursive relations become:

$$\varepsilon = \widehat{\delta}(\Delta - \varepsilon) \tag{4.117}$$

$$N(\Delta - \varepsilon) = N(\Delta) + 1 \tag{4.118}$$

The recursion begins at $\mathcal{N}(\Delta_{cr}) = 0$.

4.6.4 Least-Lifetime With Uncertain Harmonic Loads

We first consider harmonic loading with full strain reversal, in which the amplitude of the uncertain input may vary in an uncertain manner from one duty cycle to the next. Thus the sum in eq.(4.107) has only a single term:

$$f(t) = \xi \cos \frac{k_n \pi t}{T} \tag{4.119}$$

$n_f = W = 1$, and $\overline{\xi}$ and Z are known scalars. ξ is the uncertain load amplitude, which varies from one load cycle to the next between the values $\overline{\xi} - \alpha$ and $\overline{\xi} + \alpha$, according to the convex model of eq.(4.108). We will obtain an expression for the least lifetime, $N(\Delta)$, based on the assumption that the increment of damage is very small in each duty cycle.

If $\overline{\xi} \geq 0$, eqs.(4.112) and (4.113) lead to the following expression for the maximum increment of damage:

$$\widehat{\delta}(p) = \gamma Z^{\nu} \left[\overline{\xi} + \alpha\right]^{2\nu} \tag{4.120}$$

and where Z, from eq.(4.109), depends on the model parameters $p = (c, k)$, which in turn depend on the total level of damage, Δ, through eq.(4.106). The maximum damage increment, $\widehat{\delta}$, depends on the total level of damage, Δ, only through Z.

Let primes denote differentiation with respect to Δ, and approximate $N(\Delta - \varepsilon) \approx N(\Delta) - N'(\Delta)\varepsilon$. Now combining eqs.(4.117), (4.118) and (4.120) leads to:

$$N'(\Delta) = -\frac{1}{\gamma Z^{\nu} \left[\overline{\xi} + \alpha\right]^{2\nu}} \tag{4.121}$$

This relation shows how N' scales with $\overline{\xi} + \alpha$. The nominal uncertainty-free condition occurs when $\alpha = 0$. Let $N_0(\Delta)$ be the corresponding nominal lifetime. This lifetime N_0 is in fact precisely the value measured in an S–N curve. Then (4.121) indicates that:

$$N'(\Delta) = \left(\frac{\overline{\xi}}{\overline{\xi} + \alpha}\right)^{2\nu} N_0'(\Delta) \tag{4.122}$$

$N(\Delta)$ and $N_0(\Delta)$ both vanish at the critical level of damage ($N(\Delta_{cr}) = N_0(\Delta_{cr}) = 0$), so an integration on Δ results in:

$$N(\Delta) = \left(\frac{\overline{\xi}}{\overline{\xi} + \alpha}\right)^{2\nu} N_0(\Delta) \tag{4.123}$$

Thus the least lifetime curves scale inversely as the (2ν)th power of the quantity $\overline{\xi} + \alpha$. This equation establishes a relation between the lifetime (the number of duty cycles to failure) in the absence of load uncertainty, N_0, and the least lifetime consistent with the uncertainty in the load history, N. Eq.(4.123) forms the basis of our evaluation of the reliability of the vibrating system subject to fatigue failure with uncertain load history.

4.6.5 Fatigue Reliability With Uncertain Harmonic Loads

In the absence of load uncertainty, the fatigue-related damage level will reach its critical value after N_0 duty cycles. This is the laboratory S–N value of the fatigue lifetime. When the load history is uncertain, the critical damage level may be reached earlier. For the purpose of evaluating the robust reliability, we will consider the system to fail if the number of duty cycles to critical damage is less than a threshold value:

$$N \leq N_{cr} \tag{4.124}$$

The robust reliability is the least upper bound of the load uncertainty which the system can tolerate without violating condition (4.124). N_{cr} can be any value up to N_0, the number of duty cycles required to accumulate the critical level of damage in the absence of load uncertainty. When $N_{cr} < N_0$, we are considering as "acceptable", critical damage-accumulation in less than N_0 cycles.

Equating the righthand side of eq.(4.123) to the critical number of cycles to failure, N_{cr}, and solving for α leads to:

$$\widehat{\alpha} = \overline{\xi}\left[\left(\frac{N_0}{N_{cr}}\right)^{1/2\nu} - 1\right] \tag{4.125}$$

The robust reliability $\widehat{\alpha}$ is zero if $N_{cr} = N_0$, since N_0 is the number of duty cycles required to accumulate the critical amount of damage in the absence of load uncertainty. α is positive when N_{cr} is less than N_0, indicating that the system can then tolerate load uncertainty, since reaching the critical damage in more than N_{cr} cycles is not considered as "failure".

The stress level in each duty cycle varies between $\overline{\xi} - \alpha$ and $\overline{\xi} + \alpha$. Eq.(4.125) shows that the robustness index varies with the nominal load level, though not necessarily linearly since the nominal lifetime, N_0 as well as the critical lifetime, N_{cr}, may also vary with $\overline{\xi}$.

4.6.6 Fatigue Reliability With Complex Uncertain Loads

Now we consider the fatigue reliability with complex uncertain loads, represented by eqs.(4.107) and (4.108), where $n_f \geq 1$, with one simplification. We let $\overline{\xi} = 0$, which means that the nominal or "typical" load is zero. Nevertheless we can still choose one of the wave numbers k_n in eq.(4.108) to be zero, implying that during each duty cycle the stress varies around a constant value possibly different from zero. This constant value is uncertain and free to vary from one cycle to another.

The greatest possible increment of energy dissipated in a single duty cycle, eq.(4.112), is found using the method developed in section 4.5.2. This leads to an eigenvalue problem, and the result is:

$$\widehat{E}(p) = \alpha^2 \max \text{eig} \left[W^{-1/2} Z W^{-1/2} \right] \tag{4.126}$$

Let $\widehat{\lambda}$ denote the maximum eigenvalue referred to in this equation. Combining this with eq.(4.113), the greatest increment of damage in a duty cycle is:

$$\widehat{\delta}(p) = \gamma \widehat{\lambda}^{\nu} \alpha^{2\nu} \tag{4.127}$$

which is analogous to eq.(4.120). Similarly, the analog of eq.(4.121) is:

$$N'(\Delta) = -\frac{1}{\gamma \widehat{\lambda}^{\nu} \alpha^{2\nu}} \tag{4.128}$$

Now, let $N_0'(\Delta)$ be the least number of duty cycles to failure at a reference level of uncertainty, α_0. Using eq.(4.128) this can be expressed:

$$N_0'(\Delta) = -\frac{1}{\gamma \widehat{\lambda}^{\nu} \alpha_0^{2\nu}} \tag{4.129}$$

Combining eqs.(4.128) and (4.129) leads to the analog of eq.(4.122):

$$N'(\Delta) = \left(\frac{\alpha_0}{\alpha} \right)^{2\nu} N_0'(\Delta) \tag{4.130}$$

An integration leads to:

$$N(\Delta) = \left(\frac{\alpha_0}{\alpha} \right)^{2\nu} N_0(\Delta) \tag{4.131}$$

Equating the right hand side of this relation to the critical number of cycles, N_{cr}, leads to the following expression for the robustness index:

$$\widehat{\alpha} = \alpha_0 \left(\frac{N_0}{N_{cr}} \right)^{1/2\nu} \tag{4.132}$$

If the least lifetime is N_0 duty cycles with load uncertainty α_0, then $\widehat{\alpha}$ is the greatest tolerable uncertainty while requiring that the critical damage level not be reached in less than N_{cr} cycles.

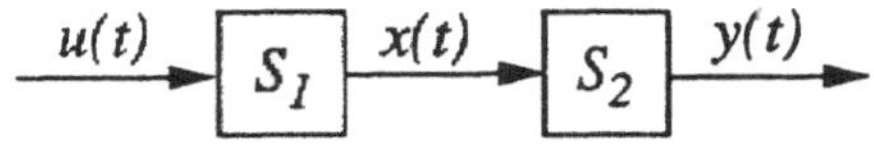

Figure 4.8: Two-unit serial network for problem 4.

4.7 Problems

1. Consider the one-dimensional undamped linear oscillator defined in eq.(4.1) and the failure set of eq.(4.5). Determine the input, failure, and overall reliabilities for the following input uncertainty models. (a) The instantaneous energy-bound convex model centered at the origin:

$$\mathcal{U}(\alpha_i) = \left\{u(t) : \ u^2(t) \le \alpha_i^2\right\} \tag{4.133}$$

(b) The Fourier ellipsoid-bound convex model of eqs.(4.50) and (4.51).

2. Consider the one-dimensional mass-spring system of eq.(4.1), driven by an input whose uncertainty is represented by a Fourier ellipsoid-bound convex model as in eqs.(4.50) and (4.51). The system fails if its cumulative elastic energy in a duration T exceeds the critical value E_{cr}. That is, the failure criterion is:

$$\frac{k}{2}\int_0^T x_u^2(t)\,\mathrm{d}t \ge E_{\text{cr}} \tag{4.134}$$

Derive an expression for the robust reliability.

3. Consider the overall reliability of the one-dimensional mass-spring system of eq.(4.1), with input and failure uncertainties of eqs.(4.2) and (4.5). Show that $\widehat{\alpha}$ increases with stiffness, k, when $\omega t \gg 0$. Explain.

4. Consider the two-unit serial network in fig. 4.8. Each unit is a mass-spring system with mass m and stiffness k. The input to S_1 is uncertain, and represented by the cumulative energy-bound convex model of eq.(4.2). The output of S_1 is the input to S_2. Both systems start from zero initial conditions. Derive an expression for the robust reliability if the network fails when the output of S_2 exceeds the critical value y_{cr}.

5. Consider the symmetric two-mass system in fig. 4.9. The forces applied to each mass are $u_1(t)$ and $u_2(t)$, and the resulting displacements from equilibrium are $x_1(t)$ and $x_2(t)$. The input and failure uncertainties are represented by the following convex models:

$$\mathcal{U}(\alpha_i) = \left\{u(t) : \ u^T(t)u(t) \le \alpha_i^2\right\} \tag{4.135}$$

$$\mathcal{F}(\alpha_f) = \left\{f(t) : \ \left[f(t) - \overline{f}\right]^T \left[f(t) - \overline{f}\right] \le \alpha_f^2\right\} \tag{4.136}$$

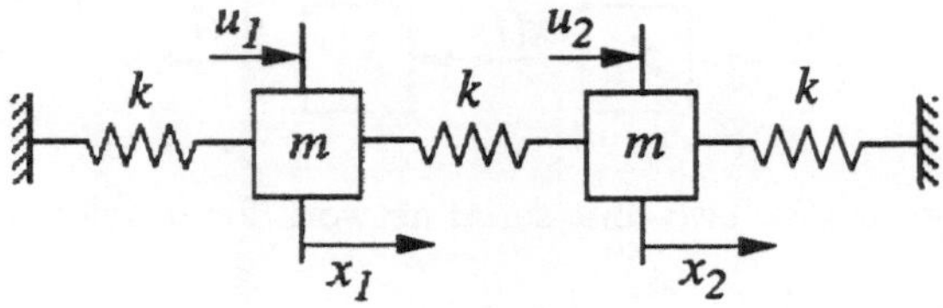

Figure 4.9: Symmetric two-mass oscillatory system for problem 5.

(a) Derive the overall reliability of the system. (b) What is the relative reliability of the two degrees of freedom, x_1 and x_2? (Hint: use hyperplane separation of the response and failure sets.) (c) The natural modes of motion are coherent and anti-coherent motion of the two masses. The eigenvectors are:

$$v_1 = (1,\,1)^T, \quad v_2 = (1,\,-1)^T \tag{4.137}$$

What is the relative reliability of these two modes? (Hint: use part (b) and the orthogonal transformation between modal coordinates and the coordinates x_1 and x_2.)

6. Use the concept of the reliability of a serial network, developed in chapter 3, to provide an interpretation of eq.(4.101).

7. ‡*Modal reliability in shell buckling.* (a) Extend example 1 by developing an expression for the reliability of the nth axial buckling mode. (b) Repeat (a) where now the convex model of eq.(4.88) is an ellipsoid defined by a diagonal matrix W. (c) The matrix W expresses the relative uncertainties in the various buckling modes. These uncertainties can be reduced in manufacture by controlling the uncertainties. What is the optimal, or most efficient, quality specification for the uncertainties of the modes?

8. *Reliability of a cart-lifting device.* Fig. 4.10 shows a device for lifting a self-propelled cart of mass m which moves along the support from C to B. A mechanism at joint B keeps the support arm BC continuously horizontal. A storage battery supplies power to an electric motor providing torque $T(t)$ at joint A to balance the moment of force $M(t)$ due to the weight of the moving cart. The mass of the support is negligible. A feedback loop regulates the current to keep the total moment at A equal to zero. Thus the horizontal arm rises at a constant angular velocity $\dot{\theta} = \omega$ radians/s. However, the total work which the torque motor can perform is limited to the energy E_{cr} in the storage battery. The cart progresses along the horizontal support with some uncertainty: the distance of the cart from point C is $s(t) = v_0 t + \sigma(t)$, where v_0 is known

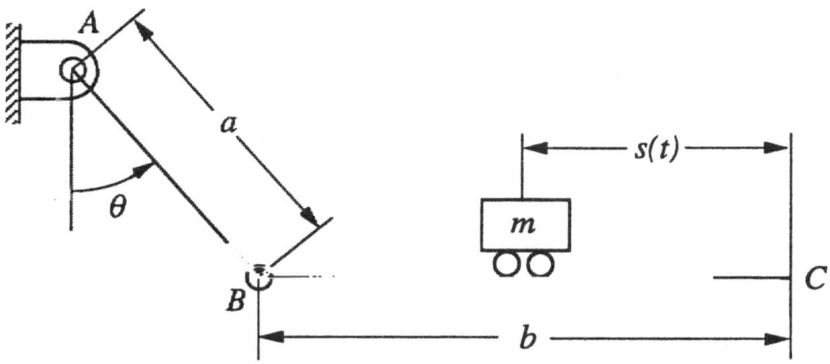

Figure 4.10: Cart-lifting device for problem 8.

and constant but $|\sigma(t)| \le \alpha$. The lifting device "fails" if the energy E_{cr} is consumed before the horizontal support has been lifted from $\theta = 0$ to $\theta = \theta_f$ radians. Calculate the robust reliability of the device.

9. *Fatigue reliability of a milling tool.* A high-speed milling tool rotates against a hard metal work piece. The lateral bending loads on the cutting tool are far below the yield strength of the material, but lead to gradual fatigue, with development of microcracks which can develop rapidly when a sudden large load is applied. For the material in question, the empirical S–N curve is:

$$\log \overline{\xi} = -0.1 \log N_0 + 6.9 \tag{4.138}$$

where N_0 is the number of pure harmonic duty cycles of amplitude $\overline{\xi}$ required to develop the critical crack density. (The logarithms are to base 10). For the same material, the energy exponent in eq.(4.105) is $\nu = 5$. Show that the robustness index, $\widehat{\alpha}$, decreases linearly with the load amplitude, if the load cycles are harmonic but with uncertain and variable amplitude.

10. *Torque excursions in an impulse turbine.* An incompressible flowing fluid does work on the rotor blades of a turbine shaft. In steady flow conditions and with some additional assumptions, the torque on the shaft resulting from the fluid load is [82, pp.622–625]:

$$M = -2\rho R A v (v \cos \beta - u) \tag{4.139}$$

where ρ is the fluid density, R is the radial distance out to the center of the fluid jet, A is the cross-sectional area of the jet, v is the input fluid speed, u is the blade speed and β is the angle between the input fluid velocity vector and the blade velocity vector. During abrupt transients in the flow velocity the blade speed will remain nearly constant due to the inertia of the turbine, while the torque exerted on the shaft can be quite large. Suppose the velocity transients are uniformly bounded, so their uncertainty is represented by a uniform-bound convex model:

$$\mathcal{V}(\alpha) = \{v(t) : \ |v(t) - \overline{v}| \leq \alpha\} \tag{4.140}$$

The turbine will operate safely provided that the torque does not exceed a critical value:

$$M \leq M_{\text{cr}} \tag{4.141}$$

What is the robust reliability of the turbine?

11. ‡ *Modification of problem 10.* The rate at which work is done on the impulse turbine of problem 10 is:

$$\frac{dW}{dt} = 2\rho A u v (v \cos\beta - u) \tag{4.142}$$

The fluid velocity is normally constant at the value $\overline{v}$, and the corresponding work done in a duration T is W_{nom}. The turbine is designed to tolerate load overshoots, provided they are not too severe, as measured by the work done in excess of the nominal value in a duration T. In other words, the failure criterion is:

$$W(T) - W_{\text{nom}} \geq W_{\text{cr}} \tag{4.143}$$

Express the uncertainty in the velocity transients with a cumulative energy-bound convex model:

$$\mathcal{V}(\alpha) = \left\{ v(t) : \int_0^{\infty} (v(t) - \overline{v})^2 \, dt \leq \alpha^2 \right\} \tag{4.144}$$

Assume the blade speed is constant during load transients. What is the robust reliability of the turbine?

Chapter 5

Fault Diagnosis, System Identification and Reliability Testing

The analysis of reliability assists the designer to make rational decisions for optimizing the performance of his system. However, even in the best of circumstances, not everything can be planned or anticipated, and the most carefully designed device may go awry if left unattended. Fault diagnosis and the monitoring of system integrity are essential for reliable operation.

In this chapter we will briefly touch on the field of system monitoring and evaluation. We will concentrate on the following three tasks:

- *Diagnosis of anomalous modifications of the system.* Cracking, breakage, deformation; these are all "failures" in the ordinary sense, and the detection and diagnosis of these and other changes is often critical for the graceful recovery or shutdown of the system as well as for subsequent repair.

- *Diagnosis of anomalous inputs.* Much of our attention in the analysis of reliability has been directed to uncertainty in the loads applied to the system. We have carefully developed the robust reliability analysis of these uncertainties, and the result is the ability to assure the performance of the system for load-uncertainty of a particular class and magnitude. In monitoring the system it is necessary to ascertain that the inputs indeed conform to the characteristics upon which the reliability analysis was performed.

- *Evaluation of the performance of the diagnostic procedure.* Anyone can formulate a diagnostic algorithm; the tough question is: how good is it? How much better is one algorithm than another? Can a particular algorithm be substantially improved, or it is essentially the best of its kind?

A multitude of approaches to these tasks is available and it is far beyond the scope of this book to review them all; we will not even try. However, the concepts of robust reliability are pertinent to these questions, and we will make a start in answering them. Our goal is to emphasize the importance of diagnostic monitoring, and to demonstrate the application of robust reliability in the formulation and evaluation of diagnostic algorithms.

5.1 Benchmark Diagnostic Resolution: Simple Examples

In this section we will use a simple example to introduce three basic concepts: fractional resolution, benchmark distinguishability and diagnostic reliability. A static load is distributed in an uncertain manner over a portion of a beam. Measurements are made to determine the extent of the load. Due to the uncertainty in the load profile, the load-extent cannot be unequivocally determined. The fractional resolution is the smallest fractional change in load-size which can be unambiguously detected. The fractional resolution is a 'benchmark' property of the measurement design, a reference point indicating the best diagnostic capability inherent in the measurement, like a benchmark reference point in land surveying. The robust reliabilty of the diagnosis is the greatest value of the load-profile uncertainty parameter consistent with an acceptable value of the fractional resolution.

5.1.1 Formulation

Structures are sometimes designed to bear substantial loads over restricted portions of their surface. This is true, for example, of structures with local supports, like cantilevers or simply-supported beams, where loads near a support can be substantially greater than away from them. Consider the idealized case of the simply supported uniform beam in fig. 5.1, which carries an uncertain load $\phi(x)$ distributed over an unknown length y. Let us assume that the load does not extend beyond the middle of the beam, so $y \le L/2$. We will measure the bending moment at the midpoint of the beam, perhaps by measuring the curvature, and we wish to determine the size of the loaded area.

The purpose of the measurement is to verify that the extent of the load is not excessive, in order to assure the ability of the beam to sustain the load. Suppose for a moment that we know the shape of the load profile, $\phi(x)$, and that the load extends from $x = 0$ to $x = y$, but that we do not know the value of y except that $y < L/2$. Measuring the bending moment at the midpoint would determine the value of y. However, the load profile $\phi(x)$

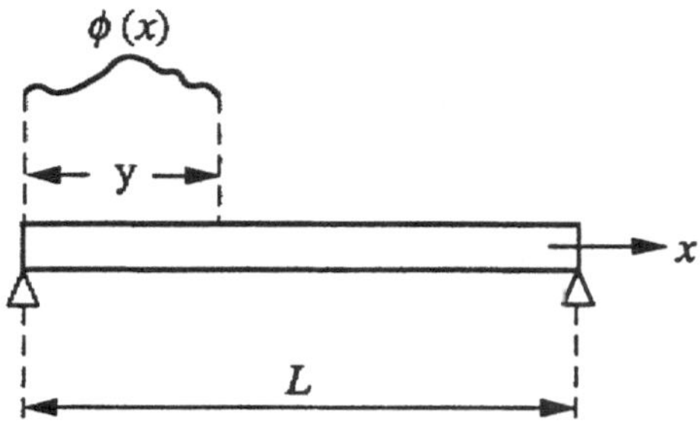

Figure 5.1: Beam with uncertain load of unknown area.

is in fact uncertain, so we will not be able to determine the precise extent of the load. Consequently, we are interested in the resolution with which we can determine y based on measuring the moment at $x = L/2$, despite the uncertainty in $\phi(x)$. In particular, we will determine the smallest fractional change in y which can be unequivocally established with this measurement. Let us denote this fractional resolution by f_y.

If the fractional resolution of y is sufficiently small, then we are satisfied with this single measurement for the purpose of establishing the safety of the beam. Let us denote the greatest acceptable fractional resolution by f_{cr}. The diagnostic procedure is acceptable if:

$$f_y \le f_{\mathrm{cr}} \tag{5.1}$$

We will find that the resolution of y depends on the amount of uncertainty in the load profile: $f_y = f_y(\alpha)$ where α is the uncertainty parameter of the load profile. The second part of the analysis is the determination of the robust reliability of the measurement: the greatest amount of uncertainty consistent with obtaining an acceptable value of the resolution. The reliability of the diagnosis assesses the performance of the measurement, and enables comparison of alternative measurement designs such as various sensor locations or number of measurements.

So, we must first determine the fractional resolution, $f_y(\alpha)$. Then we must determine the greatest uncertainty, $\widehat{\alpha}$, for which $f_y(\alpha)$ does not exceed the critical value, f_{cr}.

We will assume that the load uncertainty is specified by an envelope-bound convex model:

$$\mathcal{U}(\alpha, y) = \left\{ \phi(x) : \begin{array}{ll} |\phi(x) - \overline{\phi}| \le \alpha, & 0 \le x \le y \\ \phi(x) = 0, & y < x \le L \end{array} \right\} \tag{5.2}$$

where $\overline{\phi}$ is the positive nominal load profile and $y \le L/2$. The value of the uncertainty parameter α is unknown, but for simplicity let us assume that $\alpha < \overline{\phi}$.

5.1.2 Single Measurement

For a given value of the length y of the loaded region and a specific load profile $\phi(x) \in \mathcal{U}(\alpha, y)$, the bending moment which will be measured at the midpoint is:

$$M_\phi(L/2) = \frac{1}{2}\int_0^y x\phi(x)\,\mathrm{d}x \tag{5.3}$$

Since $\phi(x)$ is uncertain, even with fixed y, $M_\phi(L/2)$ can vary over a range of values, which we refer to as the *response set:*

$$\mathcal{R}(\alpha, y, L/2) = \{M_\phi(L/2), \quad \text{for all} \quad \phi(x) \in \mathcal{U}(\alpha, y)\} \tag{5.4}$$

This set expresses the uncertainty in the measured moment resulting from the uncertainty in the shape of the load profile. It is from this uncertainty in M_ϕ that we must establish the relative resolution of y.

The response set $\mathcal{R}(\alpha, y, L/2)$ is simply an interval: all the values from the minimum to the maximum which M_ϕ can attain. These extrema occur when the load is extremal throughout its range: $\phi(x) = \overline{\phi} \pm \alpha$. Recalling that $\alpha < \overline{\phi}$, the extremal moments are:

$$\max_{\phi(x)\in\mathcal{U}(\alpha,y)} M_\phi(L/2) = \frac{(\overline{\phi}+\alpha)y^2}{4} \tag{5.5}$$

$$\min_{\phi(x)\in\mathcal{U}(\alpha,y)} M_\phi(L/2) = \frac{(\overline{\phi}-\alpha)y^2}{4} \tag{5.6}$$

So, the response set is the interval:

$$\mathcal{R}(\alpha, y, L/2) = \frac{y^2}{4}\,[\overline{\phi}-\alpha,\ \overline{\phi}+\alpha] \tag{5.7}$$

We can always distinguish two values of the load-size, y_2 from y_1, if and only if their response sets are disjoint. For $y_2 > y_1$, the response set $\mathcal{R}(\alpha, y_2)$ is to the right of $\mathcal{R}(\alpha, y_1)$. So, these response sets are disjoint if the upper limit of $\mathcal{R}(\alpha, y_1)$ is less than the lower limit of $\mathcal{R}(\alpha, y_2)$:

$$\frac{y_1^2}{4}(\overline{\phi}+\alpha) < \frac{y_2^2}{4}(\overline{\phi}-\alpha) \tag{5.8}$$

This implies that these load-sizes are distinguishable for all load profiles $\phi(x)$ if and only if:

$$\frac{y_2}{y_1} > \sqrt{\frac{\overline{\phi}+\alpha}{\overline{\phi}-\alpha}} \tag{5.9}$$

So, the fractional resolution of y is:

$$f_y(\alpha) = \frac{y_2 - y_1}{y_1} = \sqrt{\frac{\overline{\phi}+\alpha}{\overline{\phi}-\alpha}} - 1 \tag{5.10}$$

This is the fractional resolution of the size of the loaded region. A small value of f_y means that the size of the load can be determined very accurately, despite the uncertainty in $\phi(x)$. The larger the value of f_y, the poorer the resolution. When the load uncertainty α is small, the fractional resolution is small; as α approaches $\overline{\phi}$ the fractional resolution increases without bound.

The reliability of the diagnosis of y is the greatest value of the uncertainty consistent with the required fractional resolution of the size, $f_{\rm cr}$. That is, as in the robust analysis of the reliability of a mechanical system, we equate $f_y(\alpha)$ to $f_{\rm cr}$ and solve for α:

$$f_y(\widehat{\alpha}) = f_{\rm cr} \quad \Longrightarrow \quad \widehat{\alpha} = \overline{\phi}\,\frac{(1+f_{\rm cr})^2-1}{(1+f_{\rm cr})^2+1} \tag{5.11}$$

The reliability increases monotonically from zero when the acceptable resolution vanishes, to $\overline{\phi}$ when $f_{\rm cr}$ goes to infinity.

Before we examine some variations on this example, let us indicate more generally what we have done in determining the fractional resolution of the load area. The response set $\mathcal{R}(\alpha, y)$ expands and contracts as the load-size y increases and decreases, respectively. Two load sizes y_1 and y_2 are unequivocally distinguishable for any load profile $\phi(x)$, on the basis of the measurement, if and only if the corresponding response sets are disjoint:

$$\mathcal{R}(\alpha, y_1) \cap \mathcal{R}(\alpha, y_2) = \emptyset \tag{5.12}$$

The fractional resolution is obtained from the condition of tangency of the response sets.

The fractional resolution calculated in this way is a *benchmark* value, indicating the inherent diagnostic potential of the measurement. Any pair of load sizes y_1 and y_2 whose fractional difference is less than f_y, are always distinguishable, for any load profiles in the convex model. On the other hand, a pair of load sizes whose fractional difference exceeds f_y will be indistinguishable for some allowed load profiles. No amount of "interpretation" of the measurement will be able to unequivocally differentiate between them in all cases. This type of benchmark diagnostic resolution occurs widely in fault diagnosis and in other types of measurements [6].

5.1.3 Variable Measurement Position

It may not be convenient or feasible to measure the bending moment at the midpoint of the beam, so we will now consider locating the measurement elsewhere on the beam. The magnitude of the moment will vary depending on where we measure. The question however is, do the resolution and reliability vary with measurement position ξ, for $[L/2 \le \xi < L]$? We cannot measure precisely at the end of the beam, for there the moment vanishes. Similarly, the moment near the end will be small and we should expect problems with the diagnosis due to measurement inaccuracy. We will first ignore this, and later include measurement error.

No Measurement Error. The internal bending moment at a position $\xi \geq L/2$ resulting from a load profile in $\mathcal{U}(\alpha, y)$ is:

$$M_\phi(\xi) = \left(1 - \frac{\xi}{L}\right) \int_0^y x\phi(x)\,\mathrm{d}x \tag{5.13}$$

(We assume, as before, that $y \leq L/2$.) This is simply $M_\phi(L/2)$ of eq.(5.3) multiplied by a factor. The extremal moments and the response set are therefore just multiplied by the same factor. Instead of eq.(5.7), the response set becomes:

$$\mathcal{R}(\alpha, y, \xi) = \frac{y^2}{2}\left(1 - \frac{\xi}{L}\right)\left[\overline{\phi} - \alpha,\ \overline{\phi} + \alpha\right] \tag{5.14}$$

Arguing as in eqs.(5.8)–(5.10) we see that the fractional resolution is the same for any measurement position within the interval $[L/2,\ L)$. Consequently, the reliability of the measurement will not change either.

Measurement Error. We now consider measurement error. Various measurement-error models are conceivable; let us suppose that the measured moment could vary due to instrumental error by as much as $\pm A$, regardless of the measured value. To determine the fractional resolution of the load size, we must establish the condition for disjointness of the response sets in light of both the load-profile uncertainty and the measurement error. If $y_2 > y_1$, the disjointness of $\mathcal{R}(\alpha, y_2)$ and $\mathcal{R}(\alpha, y_1)$ requires that the greatest possible measurement from the latter set be less that the least measurement from the former. Including possible measurement errors, which we take as unfavorably as possible, this condition becomes:

$$\frac{y_1^2}{2}\left(1 - \frac{\xi}{L}\right)(\overline{\phi} + \alpha) + A < \frac{y_2^2}{2}\left(1 - \frac{\xi}{L}\right)(\overline{\phi} - \alpha) - A \tag{5.15}$$

So, the fractional resolution of the load-size is:

$$f_{y,\xi} = \frac{y_2 - y_1}{y_1} \tag{5.16}$$

$$= -1 + \sqrt{\frac{\overline{\phi} + \alpha}{\overline{\phi} - \alpha} + \frac{4A}{\left(1 - \frac{\xi}{L}\right)^2 \left(\overline{\phi}^2 - \alpha^2\right) y_1^2}} \tag{5.17}$$

Due to the term $\left(1 - \frac{\xi}{L}\right)^2$ in the denominator, this fractional resolution will become quite poor — very large — when the measurement is located near the end of the beam. Furthermore, if the measurement is close to the end of the beam, so that $\left(1 - \frac{\xi}{L}\right)^2$ is large, then the fractional resolution may exceed the critical value regardless of the magnitude of the uncertainty parameter α. In other words, as the measurement position ξ moves towards L, the resolution deteriorates due to measurement uncertainty even without uncertainty in the load profile. In the absence of measurement inaccuracy, $A = 0$, and eq.(5.17) reverts to eq.(5.10) as expected, since then the fractional resolution is independent of the sensor location.

5.1.4 Multiple Measurements

From eq.(5.13) we learn that bending moments measured at different positions on the righthand side of the beam, outside the domain of the load, are simply scaled versions of one another. Multiple measurements of this sort can add no new information (ignoring the possibility of measurement noise). However, if we are able to measure one moment outside the loaded region and one within it, perhaps by embedding strain gauges, the situation is quite different, as we will see.

Let us suppose that we can measure the bending moment at two points on the beam, one inside and the other outside the domain of the load, which is of length y:

$$\xi_1 < y \quad \text{and} \quad \xi_2 > y \tag{5.18}$$

The bending moments at these points can be expressed as:

$$M(\xi_1) = \int_0^{\xi_1} \left(1 - \frac{\xi_1}{L}\right) x\phi(x)\,\mathrm{d}x + \int_{\xi_1}^{y} \xi_1 \left(1 - \frac{x}{L}\right) \phi(x)\,\mathrm{d}x \tag{5.19}$$

$$M(\xi_2) = \int_0^{y} \left(1 - \frac{\xi_2}{L}\right) x\phi(x)\,\mathrm{d}x \tag{5.20}$$

We may think of these two measurements as the elements of a vector, $M = (M_1, M_2)^T$. The response set, $\mathcal{R}(\alpha, y)$, is now not an interval like eqs.(5.7) and (5.14). Instead, the response set is a region on the M_2-versus-M_1 plane. Two load-sizes, y_1 and y_2, are always unequivocally distinguishable on the basis of this measurement vector if their response sets are disjoint. That is, if eq.(5.12) holds true. If the response sets $\mathcal{R}(\alpha, y_1)$ and $\mathcal{R}(\alpha, y_2)$ are disjoint, then whatever the load profile may be, we can never confuse y_1 and y_2, since their measurement vectors are never the same. On the other hand, if $\mathcal{R}(\alpha, y_1)$ and $\mathcal{R}(\alpha, y_2)$ have some element in common, then when that measurement vector is obtained, one can not rationally decide which load-size prevails. Disjointness of response sets is the condition for *benchmark distinguishability* of load sizes y_1 and y_2. Like the benchmark value of fractional resolution discussed on p.101, the benchmark distinguishability expresses the best resolution which is inherent in the measurement itself. No manner of data reduction or intepretation will be able to unequivocally distinguish all occurrences of load sizes which are not benchmark distinguishable, while values which are benchmark distinguishable can in principle always be separated.

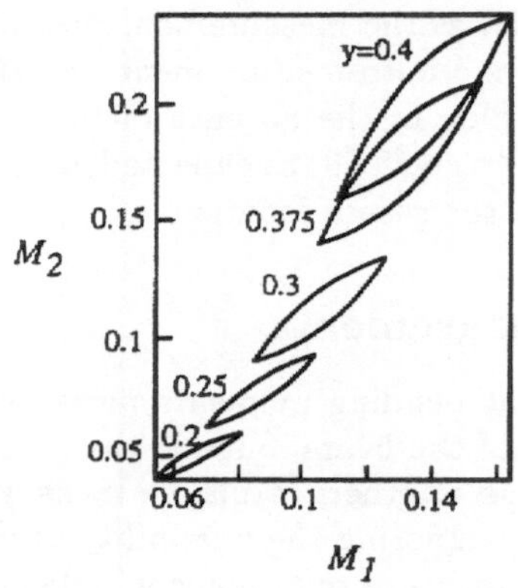

Figure 5.2: Two-dimensional response sets, $\mathcal{R}(\alpha, y)$.

So how much better are two measurements than one? Fig. 5.2 shows the outlines of response sets for five different values of load size. These sets are of course convex, so they are defined by their boundaries. In all cases the measurement locations are $\xi_1 = 0.1$ and $\xi_2 = 0.5$. The beam length is $L = 1$, and the parameters of the convex model are $\overline{\phi} = 5$ and $\alpha = 1$. The five values of the load size are $y = 0.2$ at the lower left, and take the values 0.25, 0.3, 0.375 and 0.4 going progressively to the upper right.

Consider first $y = 0.25$ and $y = 0.3$. The response sets are disjoint, so these load-sizes are always distinguishable on the basis of the pair of measurements, M_1 and M_2; these load-sizes are benchmark distinguishable with two measurements. Can either of these measurements alone, M_1 or M_2, provide unequivocal, benchmark distinguishability of these load sizes? Projecting a 2-dimensional response set onto one of the axes produces the response set for that single measured moment. Projecting $\mathcal{R}(\alpha, 0.25)$ and $\mathcal{R}(\alpha, 0.3)$ onto either the M_1 or the M_2 axis results in overlapping 1-dimensional response sets, violating the disjointness condition for benchmark distinguishability, eq.(5.12). So, neither measurement alone provides unequivocal distinguishability of these load sizes, while two measurements do.

But this happy situation does not always prevail, as in the case of the large sizes, $y = 0.375$ and $y = 0.4$. These response sets show a healthy overlap, so these load sizes are not invariably distinguishable with these two measurement locations.

5.1.5 Hyperplane Separation

The diagnostic performance of a single measurement is succintly expressed by the fractional resolution, eq.(5.10) without noise or eq.(5.17) with noise. With two measurements the situation is more complicated. From fig. 5.2 we can determine whether or not particular values of the load-size are benchmark distinguishable, but it does not give an evaluation of the performance for arbitrary load-size.

With just two measurements it would be feasible to graphically determine the resolution for each load-size. For instance, $\mathcal{R}(\alpha, 0.25)$ is disjoint from $\mathcal{R}(\alpha, 0.3)$, implying that these load-sizes are benchmark distinguishable. One could plot response sets for values of y_2 between 0.25 and 0.3 and compare them against $\mathcal{R}(\alpha, 0.25)$ until a response set is found which is tangent to $\mathcal{R}(\alpha, 0.25)$. In this way one determines the benchmark fractional resolution at $y = 0.25$, and similarly at other values of y.

This procedure becomes more and more cumbersome as the number of measurements increases. The response sets have the same dimension as the number of measurements, and their graphical representation and comparison is awkward at high dimension.

However, the method of hyperplane separation discussed in chapter 2 is readily applied here. The response sets are closed and bounded, so the disjointness condition for benchmark distinguishability, eq.(5.12), holds if and only if there is a hyperplane which separates them. A family of hyperplanes is specified by a vector w to which they orthogonal. Response sets $\mathcal{R}(\alpha, y_1)$ and $\mathcal{R}(\alpha, y_2)$ are disjoint if and only if there is a vector w such that:

$$\max_{r\in\mathcal{R}(\alpha,y_1)} w^T r < \min_{s\in\mathcal{R}(\alpha,y_2)} w^T s \tag{5.21}$$

For the particular example in question here, r and s are 2-dimensional vectors whose elements are given in eqs.(5.19) and (5.20). The inner product $w^T r$ can be presented explicitly as:

$$\begin{aligned} w^T r &= w_1 M(\xi_1) + w_2 M(\xi_2) && (5.22)\\ &= \int_0^{\xi_1} c_1 x \phi(x)\,\mathrm{d}x + \int_{\xi_1}^{y_1} c_2(x)\phi(x)\,\mathrm{d}x && (5.23)\end{aligned}$$

where we have defined:

$$\begin{aligned} c_1 &= w_1\left(1 - \frac{\xi_1}{L}\right) + w_2\left(1 - \frac{\xi_2}{L}\right) && (5.24)\\ c_2(x) &= w_1\xi_1\left(1 - \frac{x}{L}\right) + w_2 x\left(1 - \frac{\xi_2}{L}\right) && (5.25)\end{aligned}$$

The convex model of eq.(5.2) allows $\phi(x)$ to vary freely between $\overline{\phi} - \alpha$ and $\overline{\phi} + \alpha$. Consequently, $w^T r$ is maximized by choosing $\phi(x)$ to switch between these extreme values when the integrand changes sign. For a maximum we choose $\phi(x) = \overline{\phi} + \alpha$ when the integrand is positive, and $\phi(x) = \overline{\phi} - \alpha$ when the integrand is negative. The result is:

$$\max_{r\in\mathcal{R}(y_1)} w^T r = \int_0^{\xi_1} c_1 x\left(\overline{\phi} + \operatorname{sgn}(c_1)\alpha\right)\,\mathrm{d}x + \int_{\xi_1}^{y_1} c_2(x)\left(\overline{\phi} + \operatorname{sgn}[c_2(x)]\alpha\right)\,\mathrm{d}x \tag{5.26}$$

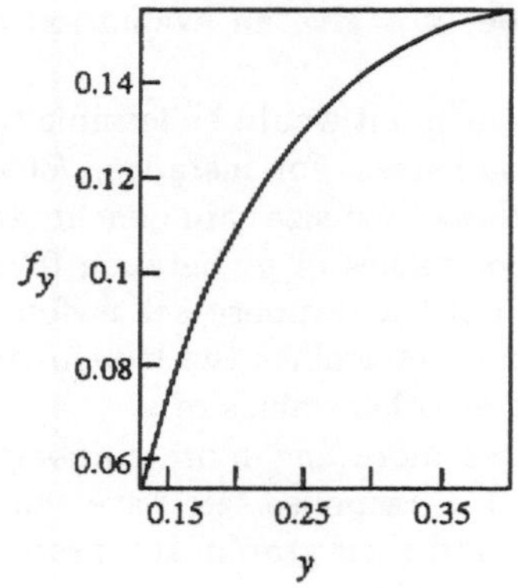

Figure 5.3: Fractional resolution versus load size, with two measurements.

$\mathrm{sgn}(c) = \pm 1$, matching the sign of c.

The minimum of $w^T s$ is constructed similarly, but by choosing the opposite direction of sign changes:

$$\min_{s \in \mathcal{R}(y_2)} w^T s = \int_0^{\xi_1} c_1 x \left(\overline{\phi} - \mathrm{sgn}(c_1)\alpha\right) \, dx + \int_{\xi_1}^{y_2} c_2(x) \left(\overline{\phi} - \mathrm{sgn}[c_2(x)]\alpha\right) \, dx \tag{5.27}$$

For any given value y_1 of the load-size, eq.(5.21) provides the basis for numerical determination of the least value y_2, greater than y_1, such that the corresponding response sets are tangent. Any value greater than y_2 is unequivocally distinguishable from y_1. In other words, $(y_2 - y_1)/y_1$ is the fractional resolution. Fig. 5.3 shows the results of this numerical determination of the fractional resolution with two measurements. The moments are measured at positions $\xi_1 = 0.1$ and $\xi_2 = 0.5$. The parameters of the convex model are $\alpha = 1$ and $\overline{\phi} = 5$.

Fig. 5.3 shows that, with two measurements, the fractional resolution varies with the size of the loaded zone. This is unlike the case of a single measurement, for which f_y is independent of y, eq.(5.10). It is noteworthy that, with a single measurement $f_y = 0.225$ for the parameter values we have used, which is substantially larger (and hence poorer) than with two measurements, as seen by comparing this value with fig. 5.3. "Two is better than one." Substantially better, and to continue with the biblical metaphor[1] we might expect continued improvement with three or more measurements, though the marginal improvement may be expected to diminish rapidly.

The curve of fig. 5.3 comprehensively demonstrates the improvement obtained by performing an additional measurement. Furthermore, one can

[1] "Two is better than one, ... and the three-fold cord will not quickly break." *Kohelet* (Ecclesiastes) 4:9–12.

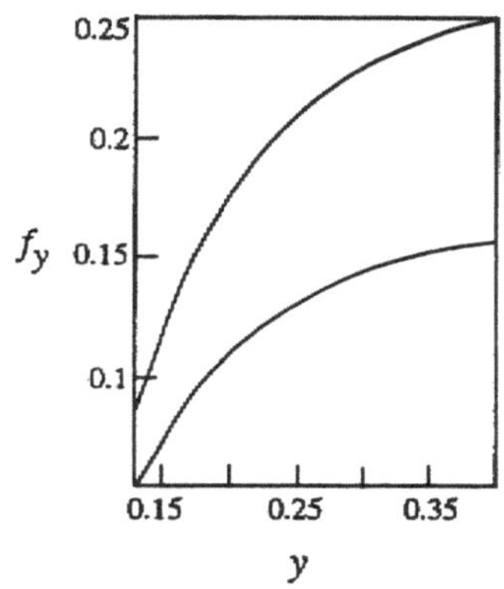

Figure 5.4: Curves of fractional resolution versus load-size, with two measurements, for $\alpha = 1$ (lower) and $\alpha = 1.5$ (upper).

compare alternative measurement locations by constructing such a curve for different values of ξ_1 and ξ_2.

5.1.6 Reliability With Two Measurements

Fig. 5.3 shows the variation of the fractional resolution versus the size of the loaded region, based on two measurements of the bending moment. These measurements, we recall, are to be used to assess the safety of the beam by determining whether or not the loaded zone is excessively large. We will accept the diagnosis if the fractional resolution of the load-size is less than a certain critical value, f_{cr}, as we did previously in connection with eq.(5.1) in section 5.1.2. If the fractional resolution is sufficiently small, we have confidence in the diagnosis.

So how reliable is the diagnosis when based on two measurements? How much uncertainty can it tolerate? How much more reliable are two measurements than only one?

The curve of fractional resolution in fig. 5.3 has been constructed for a particular value of the uncertainty parameter, $\alpha = 1$. In fig. 5.4 we reproduce this curve, together with a curve based on $\alpha = 1.5$. The measurement positions are still $\xi_1 = 0.1$ and $\xi_2 = 0.5$, and $\overline{\phi} = 5$. The layout of these curves shows, not unexpectedly, that as the uncertainty increases, the fractional resolution deteriorates. This is precisely the information needed for constructing the robust reliability of the measurement.

We reproduce these two resolution curves in fig. 5.5, together with parts of five additional curves. The dashed line shows a constant value of $f_{cr} = 0.15$. We calculate the reliability from the points of intersection of the solid fractional-resolution curves with the dashed critical-value line. For instance, the dashed line intersects the $\alpha = 1$ curve (bottom-most) at $y = 0.35$. This

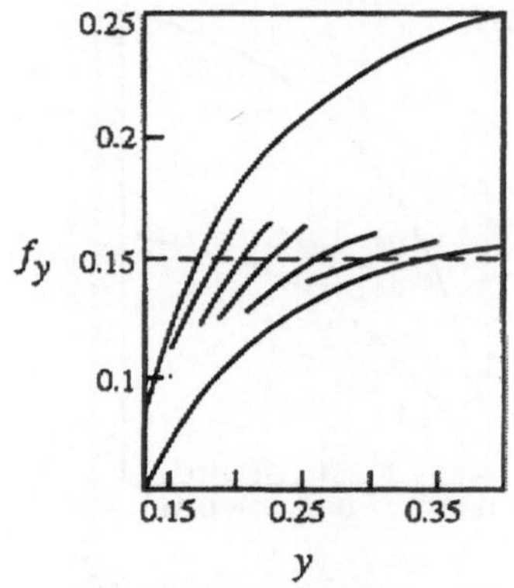

Figure 5.5: Curves of fractional resolution versus load-size, with two measurements, for $\alpha = 1$, 1.05, 1.1, 1.2, 1.3, 1.4 and 1.5 from bottom to top.

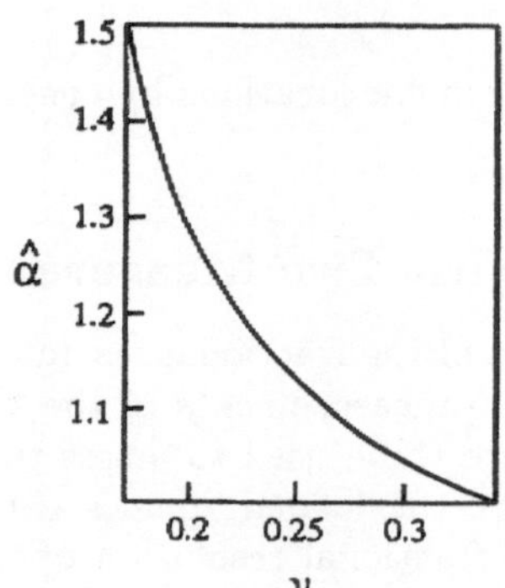

Figure 5.6: Robust reliability versus load-size, for two measurements.

means that the fractional resolution in measuring a load-zone whose size is 0.35 will exceed f_{cr} if the uncertainty parameter exceeds unity. In other words, $\alpha = 1$ is the greatest tolerable uncertainty consistent with fractional resolution no greater than $f_{cr} = 0.15$; this is the robust reliability for load-size $y = 0.35$. We determine the ordinate of intersection for each curve in this figure, and construct the robust reliability as a function of load-size. This is presented in fig. 5.6.

Fig. 5.6 indicates that load-zones whose sizes range from about $y = 0.15$ to $y = 0.4$ have robust reliability in the range of $\widehat{\alpha} = 1$ to $\widehat{\alpha} = 1.5$, when two measurements are used. From eq.(5.11) we can calculate the reliability for a single measurement, obtaining $\widehat{\alpha} = 0.69$ for $\overline{\phi} = 5$ and $f_{cr} = 0.15$. The two-measurement diagnosis can tolerate substantially more uncertainty without violating the condition for acceptability of the diagnosis; the extra measurement invests the diagnosis with considerably greater robustness to uncertainty, making it more reliable.

To summarize this section, we have evaluated the diagnosis of the size of uncertain load profiles with two different concepts: the fractional resolution and the robust reliability. The fractional resolution, f_y, indicates the uncertainty in determining the load-size which results from the uncertainty in the shape of the load profile. The fractional resolution is a "benchmark" characteristic of the system. It is evaluated from disjointess of response sets, and expresses the diagnostic potential which is inherent in the system, without consideration of any specific diagnostic-decision algorithm. The robust reliability, $\widehat{\alpha}$, is the greatest value of the load-profile uncertainty parameter for which the fractional resolution does not exceed a critical value. $\widehat{\alpha}$ measures the robustness of the diagnosis to profile uncertainty. These quantities, f_y and $\widehat{\alpha}$, are complimentary tools for assessing and optimizing the diagnostic performance. We have used them for evaluating the number and deployment of measurements.

5.2 Multi-Hypothesis Diagnosis of Anomalous Inputs

In the previous section we evaluated the diagnostic potential of an uncertainly loaded beam, without actually considering any particular diagnostic procedure. The fractional resolution is derived from the disjointness criterion for benchmark distinguishability. It expresses a resolution-capability inherent in the measurements themselves. Now we wish to evaluate the performance of realizable diagnostic algorithms, based on testing specific hypothesized anomalies against the measurements.

5.2.1 Multi-Hypothesis Diagnosis

Let us consider a dynamic system which is excited by an unknown anomalous input vector $f(t)$. Denote the resulting measurement vector by $y_f(t)$. The aim of the diagnosis is to characterize the input. For instance, it may be necessary to verify that the structure and uncertainty of the input convex model is consistent with reliability of the system.

A common and important example is the N-dimensional proportionally damped linear elastic system:

$$M\ddot{x}(t) + C\dot{x}(t) + Kx(t) = Bu(t) + f(t) \tag{5.28}$$

where $u(t)$ is the normal (perhaps noisy) input and $f(t)$ is an unknown anomalous input. The measurement vector is:

$$y_f(t) = Hx(t) \tag{5.29}$$

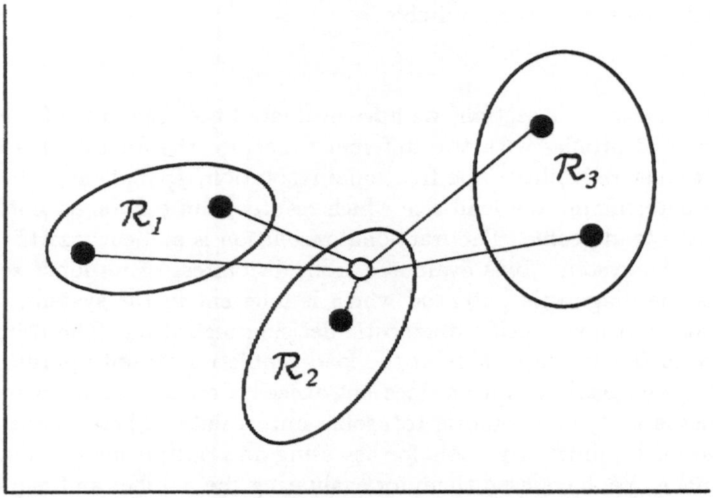

Figure 5.7: Schematic illustration of the nearest-neighbor multi-hypothesis decision rule.

The anomalous inputs $f(t)$ come from a collection of convex models, $\mathcal{F}_1(\alpha)$, $\mathcal{F}_2(\alpha)$, ... The purpose of the diagnosis is to characterize the anomaly by deciding from which convex models it could have come. We do this by choosing one or more representative elements from each input set, and comparing the measured response against the responses calculated from each of these "hypothesized" representative inputs.

Let H denote the collection of all the hypotheses chosen from all the input sets. *Multi-hypothesis diagnosis* is based on the *nearest-neighbor decision rule:* choose the hypothesized input (and hence the input set from which it came) whose calculated response is closest to the measurement. That is, given a measurement, y, we choose hypothesis $h \in H$ if:

$$||y - y_h|| = \min_{g \in H} ||y - y_g|| \tag{5.30}$$

The operation of this nearest-neighbor multi-hypothesis decision rule is illustrated schematically in fig. 5.7. Each ellipse represents a response set, $\mathcal{R}_n$: the set of all measurements obtainable from the input set $\mathcal{F}_n$. The dots (•) are calculated responses to hypothesized inputs, and the open circle (∘) is the measurement. The response sets $\mathcal{R}_1$, $\mathcal{R}_2$ and $\mathcal{R}_3$ are disjoint, so the corresponding input sets $\mathcal{F}_1$, $\mathcal{F}_2$ and $\mathcal{F}_3$ are benchmark distinguishable. That is, these three types of input anomalies can in principle always be distinguished on the basis of the measured bending moment. In addition, the measurement belongs to $\mathcal{R}_2$, indicating that it has arisen from an input in $\mathcal{F}_2$. The algorithm chooses the closest hypothesis, which in this example

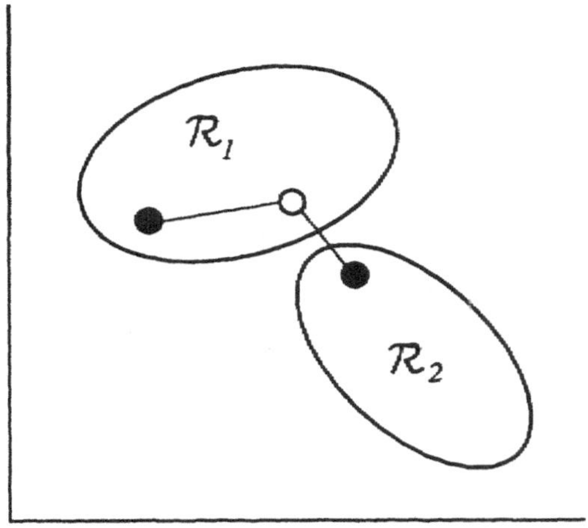

Figure 5.8: A false nearest-neighbor multi-hypothesis decision.

lies in $\mathcal{R}_2$, so the algorithm has made the correct choice by assigning the measurement to an hypothesis from the same set.

5.2.2 Criterion for Successful Diagnosis

Fig. 5.8 illustrates the possibility of false diagnosis by the nearest-neighbor rule. In this case the measurement (∘) is closest to an hypothesized response (•) which lies in a different response set. This collection of hypotheses, H, has failed to correctly diagnosis this particular input from $\mathcal{F}_1$.

It is reasonable to say that anomalous inputs from $\mathcal{F}_n$ are *correctly diagnosed* by the set of hypotheses H, if every input in $\mathcal{F}_n$ is assigned to an hypothesis from the same set. Note that this idea does not preclude the possibility that non-$\mathcal{F}_n$ inputs are also ascribed to $\mathcal{F}_n$. We will develop a criterion for deciding if an input set is correctly diagnosed, and this criterion must be applied to each input set in order to establish the global success of the algorithm.

We can now formulate a quantitative criterion for correct diagnosis of anomalous inputs [7]. Let f and g be two hypothesized inputs in the hypothesis set H. Define the *minimum relative norm* on the response set $\mathcal{R}_n$ as:

$$D_n(f,g) = \min_{y \in \mathcal{R}_n} \left[||y_f - y||^2 - ||y_g - y||^2 \right] \tag{5.31}$$

If $D_n(f,g)$ is positive, then every measured response in $\mathcal{R}_n$ is closer to hypothesized response y_g than to hypothesized response y_f. So, if $D_n(f,g)$ is positive, then every input of type n will be ascribed to hypothesis g rather than to f. If $D_n(f,g)$ is negative, then some type-n inputs will be ascribed to hypothesis f, but some may perhaps be ascribed to g. In other words, inputs from $\mathcal{F}_n$ are correctly diagnosed, in the sense defined previously, if and only if, for each hypothesis f which does not belong to $\mathcal{F}_n$, there is an hypothesis $g \in \mathcal{F}_n$ such that:

$$D_n(f,g) > 0 \tag{5.32}$$

This means that, for every input from $\mathcal{F}_n$, no hypothesis outside $\mathcal{F}_n$ will be chosen by the multi-hypothesis algorithm. Consequently type-n inputs will be correctly diagnosed.

Various vector norms can be used in eq.(5.31). The most common choice is the euclidean vector norm, $||x|| = \sqrt{x^T x}$, which we will use here.

Expanding the norms of the righthand side of eq.(5.31) in terms of the inner product, one finds:

$$D_n(f,g) = ||y_f||^2 - ||y_g||^2 - 2 \max_{y \in \mathcal{R}_n} (y_f - y_g)^T y \tag{5.33}$$

(The minimum has been expressed as "minus a maximum".)

We have reached an optimization problem with convex sets. To simplify this optimization, we exploit theorem 1 from chapter 2 (p. 21), which states that a linear function has the same extrema on a closed and bounded set and on the extreme points of that set.

If the response set $\mathcal{R}_n$ is the convex hull of a fundamental response set P_n, then the maximum in eq.(5.33) can be sought on P_n. Thus eq.(5.33) becomes:

$$D_n(f,g) = ||y_f||^2 - ||y_g||^2 - 2 \max_{y \in \mathrm{P}_n} (y_f - y_g)^T y \tag{5.34}$$

Furthermore, let the input set $\mathcal{F}_n$ be the convex hull of a fundamental input set Φ_n. Variation of y on P_n is obtained by recognizing that $y = y_\phi$ is a function of the input ϕ, and by allowing ϕ to vary on Φ_n. Thus eq.(5.34) can be written as:

$$D_n(f,g) = ||y_f||^2 - ||y_g||^2 - 2 \max_{\phi \in \Phi_n} (y_f - y_g)^T y_\phi \tag{5.35}$$

5.2.3 Example

Let us now consider the evaluation of a specific multi-hypothesis diagnosis. Continuing with the example studied in section 5.1, we will measure the bending moment at the midpoint of the beam, and attempt to determine the size of the loaded area by testing hypothesized load profiles. The load uncertainty is represented by $\mathcal{U}(\alpha, y)$ in eq.(5.2), for which the nominal load is constant at $\overline{\phi}$ throughout the loaded zone, and zero elsewhere, and we assume that $\overline{\phi} > \alpha$. We adopt the nominal load as the hypothesized profile. That is, for any size y_i, the hypothesized load profile is:

$$h_i(x) = \begin{cases} \overline{\phi}, & 0 \le x \le y_i \\ 0, & y_i < x \le L \end{cases} \tag{5.36}$$

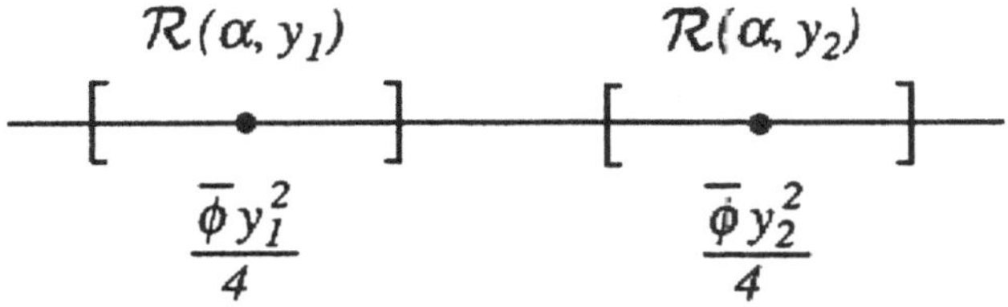

Figure 5.9: Response sets, each with an hypothesis at its midpoint.

We will determine the load-sizes y_1 and y_2 which the multi-hypothesis algorithm will always correctly distinguish, based on the hypothesized load profiles, $h_1(x)$ and $h_2(x)$.

For any two load-sizes, $y_1 < y_2$, we know that a measurement of the bending moment at the midpoint is, in principle, capable of distinguishing these values for any load profiles, if and only if $(y_2 - y_1)/y_1$ exceeds the fractional resolution, f_y, which we found in eq.(5.10). When $\overline{\phi} = 5$ and $\alpha = 1$ the fractional resolution is 0.225. Will the multi-hypothesis algorithm perform this well? Will the multi-hypothesis decision rule exploit the full potential of the measurement, as expressed by the benchmark property f_y?

We must evaluate the minimum relative norm for each of the load-profile sets:

$$D_1(h_2, h_1) = \min_{\phi \in \mathcal{U}(\alpha, y_1)} \left[(M_{h_2} - M_\phi)^2 - (M_{h_1} - M_\phi)^2 \right] \tag{5.37}$$

$$= M_{h_2}^2 - M_{h_1}^2 - 2 \max_{\phi \in \mathcal{U}(\alpha, y_1)} (M_{h_2} - M_{h_1}) M_\phi \tag{5.38}$$

$$D_2(h_1, h_2) = \min_{\phi \in \mathcal{U}(\alpha, y_2)} \left[(M_{h_1} - M_\phi)^2 - (M_{h_2} - M_\phi)^2 \right] \tag{5.39}$$

$$= M_{h_1}^2 - M_{h_2}^2 - 2 \max_{\phi \in \mathcal{U}(\alpha, y_2)} (M_{h_1} - M_{h_2}) M_\phi \tag{5.40}$$

Every load profile in $\mathcal{U}(\alpha, y_1)$ is assigned to the correct hypothesis, h_1, if and only if $D_1 > 0$. This condition assures correct diagnosis of load-size y_1, but it does not preclude the possibility that some loads of size y_2 will also be interpreted, incorrectly, as being of size y_1. Correct diagnosis of loads of size y_2 is assured if and only if $D_2 > 0$. We need to evaluate both relative norms, D_1 and D_2, if we wish to establish the complete bi-lateral correctness of the multi-hypothesis algorithm with hypotheses h_1 and h_2.

Employing eq.(5.3) for the bending moment at the midpoint we calculate the hypothesized responses to be:

$$M_{h_i} = \frac{\overline{\phi} y_i^2}{4}, \quad i = 1, 2 \tag{5.41}$$

Fig. 5.9 shows response sets $\mathcal{R}(\alpha, y_1)$ and $\mathcal{R}(\alpha, y_2)$ with the hypothesized responses at their midpoints indicated by dots ($\bullet$).

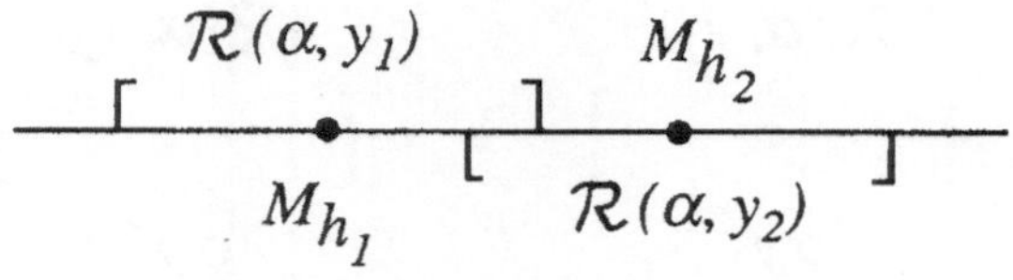

Figure 5.10: Overlapping response sets for which only y_1 is correctly diagnosed.

Using eq.(5.3) again, and eq.(5.38) for the relative norm, we can express D_1 as:

$$D_1 = \frac{\overline{\phi}^2(y_2^4 - y_1^4)}{16} - 2 \max_{\phi \in \mathcal{U}(\alpha, y_1)} \frac{\overline{\phi}(y_2^2 - y_1^2)}{8} \int_0^{y_1} x\phi(x)\,dx \tag{5.42}$$

The coefficient of the integral is positive, so the maximum occurs when $\phi(x) = \overline{\phi} + \alpha$, resulting in:

$$D_1 = \frac{\overline{\phi}^2(y_2^4 - y_1^4)}{16} - \frac{\overline{\phi}(\overline{\phi} + \alpha) y_1^2 (y_2^2 - y_1^2)}{8} \tag{5.43}$$

After some algebraic manipulations one finds that $D_1 > 0$ and y_1 is correctly diagnosed if y_1 and y_2 are related as follows:

$$y_2 > y_1 \sqrt{1 + \frac{2\alpha}{\overline{\phi}}} \tag{5.44}$$

When $\overline{\phi} = 5$ and $\alpha = 1$, we find from this relation that y_1 is correctly identified in comparison with y_2 if $(y_2 - y_1)/y_1 > 0.183$. The hypotheses of eq.(5.36) distinguish the lesser from the greater load-size with a fractional resolution which is *better* than the benchmark value of 0.225. How is this possible? When the fractional difference, $(y_2 - y_1)/y_1$, is less than the benchmark value the response sets overlap; this is the definition of the benchmark fractional resolution. This situation is illustrated in fig. 5.10. All the elements of $\mathcal{R}(\alpha, y_1)$ are closer to M_{h_1} than to M_{h_2}, so load-size y_1 is invariably correctly identified. However, some elements of $\mathcal{R}(\alpha, y_2)$ are also closer to M_{h_1} than to M_{h_2}, so load-size y_2 is not always diagnosed properly, and is sometimes identified as y_1.

Now we evaluate the reverse criterion for distinguishability, D_2, which can be expressed:

$$D_2 = \frac{\overline{\phi}^2(y_1^4 - y_2^4)}{16} - 2 \max_{\phi \in \mathcal{U}(\alpha, y_2)} \frac{\overline{\phi}(y_1^2 - y_2^2)}{8} \int_0^{y_2} x\phi(x)\,dx \tag{5.45}$$

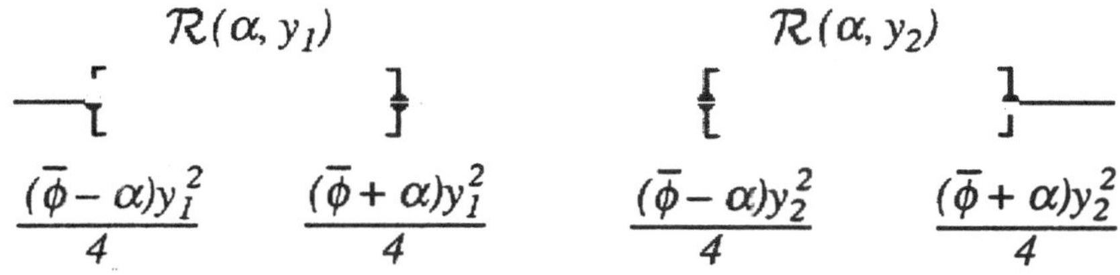

Figure 5.11: Response sets, each with two hypothesized responses.

The coefficient of the integral now is negative, so the maximum occurs when $\phi(x) = \overline{\phi} - \alpha$, resulting in:

$$D_2 = \frac{\overline{\phi}^2(y_1^4 - y_2^4)}{16} - \frac{\overline{\phi}(\overline{\phi} - \alpha)y_2^2(y_1^2 - y_2^2)}{8} \tag{5.46}$$

Again some algebraic manipulations show that $D_2 > 0$ and y_2 is correctly diagnosed if y_1 and y_2 are related as follows:

$$y_2 > y_1\sqrt{\frac{\overline{\phi}}{\overline{\phi} - 2\alpha}} \tag{5.47}$$

provided that $\overline{\phi} > 2\alpha$. When $\overline{\phi} = 5$ and $\alpha = 1$, this relation shows that y_2 is always distinguishable from y_1 if and only if their fractional difference, $(y_2 - y_1)/y_1$, exceeds 0.291.

So, with these particular hypothesized load profiles, the fractional resolution of the multi-hypothesis algorithm for correctly diagnosing y_1 is 0.183, while its fractional resolution for correctly identifying y_2 is 0.291. In other words, using this particular realization of the multi-hypothesis decision procedure, we are able to unequivocally distinguish two load sizes if and only if their fractional difference exceeds 0.291. The benchmark value, we recall, is $f_y = 0.225$, for always distinguishing any two load sizes. This multi-hypothesis algorithm is slightly poorer than the limiting performance expressed by the benchmark resolution.

Is there a multi-hypothesis algorithm whose performance is closer to the benchmark value? How much does the performance improve if we consider the use of two hypotheses for each load size, rather than only one? For example, suppose we employ two hypotheses for each set, one at each extremum, rather than a single hypothesis at the midpoint as before. That is, for the load-profile set $\mathcal{U}(\alpha, y_i)$ we test the measured moment against each of the following hypotheses:

$$h_{i,1}(x) \quad = \quad \begin{cases} \overline{\phi} - \alpha, & 0 \leq x \leq y_i \\ 0, & y_i < x \leq L \end{cases} \tag{5.48}$$

$$h_{i,2}(x) = \begin{cases} \overline{\phi}+\alpha, & 0 \le x \le y_i \\ 0, & y_i < x \le L \end{cases} \tag{5.49}$$

The responses — bending moments — calculated for these hypothesized load profiles are:

$$M_{h_{i,1}} = \frac{(\overline{\phi}-\alpha)y_i^2}{4} \tag{5.50}$$

$$M_{h_{i,2}} = \frac{(\overline{\phi}+\alpha)y_i^2}{4} \tag{5.51}$$

Two response sets with these hypothesized responses shown by dots appear in fig. 5.11. It is evident that these response sets will be successfully distinguished by this multi-hypothesis algorithm provided the sets are disjoint. In other words, the fractional resolution of this algorithm will equal the benchmark performance for two measurements. No further improvement can be achieved by altering the two hypotheses.

5.2.4 Robust Reliability

We will now study the reliability of multi-hypothesis diagnosis of input anomalies. The basic question of robust reliability is: how much uncertainty can the diagnosis tolerate without failure?

As before, a measured response is denoted y_f, where the anomalous input function $f(t)$ belongs to one or more of the collection of input sets, $\mathcal{F}_1(\alpha)$, $\mathcal{F}_2(\alpha)$, ... Representative "hypothesized" inputs, h_1, h_2, ..., are chosen from all the input sets and stored in the set H. The measured response is compared in eq.(5.30) against each of the hypothesized responses to find the nearest hypothesis, thus determining the set which is selected as responsible for the input.

The positivity of the minimum relative norm, eq.(5.32), which must hold for each input set $\mathcal{F}_n(\alpha)$, defines the condition for no-failure of the multi-hypothesis diagnostic algorithm. The robust reliability is the greatest value of the uncertainty parameter, α, consistent with this condition.

We can succintly express the robust reliability of the multi-hypothesis diagnosis as follows:

$$\widehat{\alpha} = \sup \{\alpha : \ D_n(\alpha) > 0, \quad \text{for all } n\} \tag{5.52}$$

The robust reliability is the greatest value of the uncertainty parameter which is consistent with correct diagnosis of each of the input sets. (We have written $D_n(\alpha)$ rather than $D_n(f,g)$ as previously, to stress the dependence of the minimum relative norm on the uncertainty parameter, rather than on the hypotheses.)

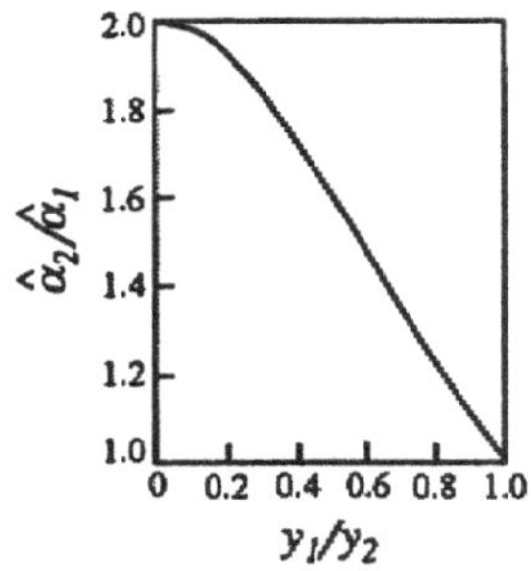

Figure 5.12: Ratio of reliabilities for two- and one-hypothesis algorithms.

Example 1 Consider the example of section 5.2.3, with a single measurement of the bending moment at the midpoint of the beam, and a single hypothesized load profile, eq.(5.36), for each input set. Eq.(5.44) is the condition for $D_1(\alpha) > 0$, while eq.(5.47) is required for $D_2(\alpha) > 0$. What is the greatest uncertainty consistent with both conditions? For D_1 we obtain:

$$D_1(\widehat{\alpha}_1) = 0 \quad \Longrightarrow \quad \widehat{\alpha}_1 = \frac{\overline{\phi}}{2}\left[\left(\frac{y_2}{y_1}\right)^2 - 1\right] \tag{5.53}$$

Any larger value of α would result in a negative value for D_1. For D_2 we find the greatest value of the uncertainty parameter is:

$$D_2(\widehat{\alpha}_2) = 0 \quad \Longrightarrow \quad \widehat{\alpha}_2 = \frac{\overline{\phi}}{2}\left[1 - \left(\frac{y_1}{y_2}\right)^2\right] \tag{5.54}$$

With the condition that $y_2 > y_1$, it results that $\widehat{\alpha}_1 > \widehat{\alpha}_2$. Consequently $\widehat{\alpha}_2$ is the greatest value of uncertainty consistent with correct diagnosis of both sets. So, the robust reliability of the multi-hypothesis diagnosis with one hypothesis for each input set is:

$$\widehat{\alpha}_{1\,\mathrm{hyp}} = \frac{\overline{\phi}}{2}\left[1 - \left(\frac{y_1}{y_2}\right)^2\right] \tag{5.55}$$

The diagnosis will sometimes fail for any greater value of uncertainty. This relation shows that the reliability of the diagnosis depends on the ratio of the load-sizes which are to be distinguished. When the load-sizes are nearly the same, and y_1 is only slightly less than y_2, the reliability is low: even small uncertainty in the load profile can lead to failure of the diagnosis. On the other hand, when $y_1/y_2 \ll 1$, greater profile-uncertainty can be tolerated. ∎

Example 2 Now consider the extension of the previous example to the case of two hypothesized load profiles for each input set, specified by eqs.(5.48) and (5.49). We found that these hypotheses will correctly diagnosis the input sets provided that the response sets are disjoint. So, the condition for correct diagnosis is simply the disjointness condition, eq.(5.8) or, equivalently, eq.(5.9):

$$\frac{y_2}{y_1} > \sqrt{\frac{\overline{\phi} + \alpha}{\overline{\phi} - \alpha}} \tag{5.56}$$

The robust reliability with this choice of hypothesized profiles is the upper bound of the values of the uncertainty parameter for which this relation holds:

$$\widehat{\alpha}_{2\,\text{hyp}} = \overline{\phi}\, \frac{1 - (y_1/y_2)^2}{1 + (y_1/y_2)^2} \tag{5.57}$$

As in the one-hypothesis case, the reliability increases as the ratio of the load-sizes decreases: small differences are less reliably diagnosed than large differences. Furthermore, comparing eqs.(5.55) and (5.57), we see that the two-hypothesis algorithm is always better than the one-hypothesis case:

$$\widehat{\alpha}_{2\,\text{hyp}} > \widehat{\alpha}_{1\,\text{hyp}} \tag{5.58}$$

for all values of $y_1 < y_2$. The ratio of $\widehat{\alpha}_{2\,\text{hyp}}$ to $\widehat{\alpha}_{1\,\text{hyp}}$ is shown in fig. 5.12. The two-hypothesis algorithm is much more reliable than the one-hypothesis case for large size-differences, $y_1/y_2 \ll 1$. At the other extreme, when distinguishing between similar load-sizes, these algorithms have nearly the same reliability. ■

5.3 Least-Squares Estimation

Many fault-diagnosis schemes are based on least-squares parameter estimation, in which an unknown vector x is determined from a set of linear algebraic equations:

$$Ax = b \tag{5.59}$$

The matrix A and the vector b are known, though perhaps subject to uncertainty. There may be a unique value of x which exactly satisfies eq.(5.59). Or these equations may be "over-determined" and contain conflicting information, or they may be "under-determined" and insufficient to fix a unique value for the vector x. In any case, a vector is sought which is small in some sense and which minimizes the squared-error of the equation. This method has a long history dating back to Gauss [46] and Legendre [91], and there are many variations, some based on the use of generalized inverses [56], or employing regularization methods [97] or other techniques [39, 42, 63, 87] to obtain a "reasonable" solution.

In this section we will examine the reliability of fault diagnosis based on a simple least-squares solution method. The purpose is primarily to illustrate several applications of robust reliability analysis. We trust the reader will be able to apply these techniques to his own problems. The linear relation (5.59) arises in a wide range of applications. We will motivate it here based on uncertain inputs to a dynamic system, and in section 5.4 we will derive a least-squares problem from modal measurements for crack diagnosis.

5.3.1 Formulation of the Least-Squares Problem

Consider a multi-dimensional linear dynamic system with state-vector $y(t)$ and uncertain scalar input $u(t)$:

$$\dot{y}(t) = By(t) + Cu(t), \quad y(0) = 0 \tag{5.60}$$

Let us represent the input in an interval $[0, T]$ by a truncated Fourier series with unknown coefficients:

$$u(t) = \sum_{k=1}^{K} x_n \cos \frac{n\pi t}{T} \tag{5.61}$$

$$= x^T \gamma(t) \tag{5.62}$$

where $\gamma(t)$ is a vector of cosine functions and x is the vector of unknown Fourier coefficients. We wish to detect anomalies in the input spectrum, x, by measuring the response, which can be written:

$$y(t) = \overbrace{\left[\int_0^t e^{B(t-\tau)} C\gamma^T(\tau)\, d\tau\right]}^{Q(t)} x \tag{5.63}$$

$$= Q(t)x \tag{5.64}$$

The matrix Q is known and can be calculated from our prior information about the system. The response y is measured and the spectrum x is sought. We may in fact measure the output at r different time instants $t_1, \ldots, t_r$ and concatenate the responses in a single vector:

$$b = \left(y^T(t_1), \ldots, y^T(t_r)\right)^T \tag{5.65}$$

$$= \underbrace{[Q(t_1), \ldots, Q(t_r)]^T}_{A} x \tag{5.66}$$

$$= Ax \tag{5.67}$$

We have now formulated a set of linear equations in the form of (5.59), which is the starting point for least-squares analysis.

Eq.(5.67) is "solved" in a least-squares sense by multiplying on the left by a generalized inverse[2] of A, denoted A^+:

$$x = A^+ b \tag{5.68}$$

[2]The generalized inverse A^+ is defined in such a way that $x = A^+b$ minimizes the norm of $Ax - b$. The simplest case arises if A^TA is non-singular, for then $A^+ = (A^TA)^{-1}A^T$.

How do the uncertainties on the righthand side propagate to the left, and how do they influence the diagnosis of anomalies in the input? How reliable is the diagnosis?

5.3.2 Variation of the Least-Squares Solution

The measurements b are uncertain, but we will suppose that A and A^+ are precisely known. In section 5.4 we will consider uncertainty in the coefficient matrix A.

Let us suppose that the measurements cluster around the correct value, $\bar{b}$, in an ellipsoid whose shape is determined by the positive definite, real, symmetric matrix W:

$$\mathcal{B}(\alpha) = \left\{b: \ (b-\bar{b})^T W (b-\bar{b}) \leq \alpha^2\right\} \tag{5.69}$$

The uncertainty parameter α determines the size of the set. In the reliability analysis of least-squares estimation we ask how large α can be without causing failure of the estimate.

One sense in which the least-squares estimate can fail is by magnifying the measurement-uncertainty into large variation of the solution itself. This is quite a common source of difficulty in least-squares estimation. The instability or 'ill-posedness' of least-squares (and other) inverse problems is a major area of research [65, 97].

The solution, x, will change as the measurement varies on the set $\mathcal{B}(\alpha)$. One can assess the stability of x as the difference between the greatest and least values of the euclidean norm of x. This involves optimizing a quadratic function — the norm — on the set $\mathcal{B}$. A simpler and often just as useful assessment is the extremal variations of the projection of x along an arbitrary direction, represented by a unit vector ψ. This requires optimizing a linear function — the projection — on $\mathcal{B}$. The projection is related to the norm as:

$$x^T\psi = ||x|| \cdot ||\psi|| \cos \angle(x,\psi) \tag{5.70}$$

where $\angle(x,\psi)$ is the angle between the two vectors. In the course of our analysis we will find the vector ψ which produces the greatest variation of the projection.

The projection defines a failure criterion: the estimate is unacceptable if the variation of the projected solution exceeds a critical value, p_{cr}:

$$\max x^T\psi - \min x^T\psi \ > \ p_{\mathrm{cr}} \tag{5.71}$$

Using relation (5.68), the projection of the solution along ψ is $\psi^T A^+ b$. So, the optimization problem we face is:

$$\text{opt } \psi^T A^+ b \quad \text{subject to} \quad b \in \mathcal{B}(\alpha) \tag{5.72}$$

$\psi^T A^+ b$ is a linear function and $\mathcal{B}$ is a closed and bounded convex set, so the extrema occur on the boundary of $\mathcal{B}$, as explained in theorem 1 of chapter 2 (p.21). Consequently, (5.72) can be written:

$$\text{opt}\ \psi^T A^+ b \quad \text{subject to} \quad \left(b-\overline{b}\right)^T W \left(b-\overline{b}\right) = \alpha^2 \tag{5.73}$$

A change of variables is useful, defining $\beta = b - \overline{b}$. Following example 5 of chapter 2 (p.21) we find the extremal values of $\psi^T x$ to be:

$$\underset{b\in\mathcal{B}(\alpha)}{\text{opt}}\ \psi^T x \ = \psi^T A^+ \overline{b} \pm \alpha\sqrt{\psi^T A^+ W^{-1} A^{+^T} \psi} \tag{5.74}$$

The range of variation of the projection is simply $2\alpha\sqrt{\psi^T A^+ W^{-1} A^{+^T}\psi}$. The projection direction is represented by ψ, which is a unit vector, so the quadratic term $\psi^T A^+ W^{-1} A^{+^T}\psi$ can be as large as the greatest eigenvalue of the positive semi-definite, real, symmetric matrix $A^+ W^{-1} A^{+^T}$. We will denote this maximal eigenvalue by λ_{max}. This extremal value is obtained when ψ is proportional to the corresponding eigenvector [56]. So, the greatest possible variation of *any* projection of x is:

$$\max_{\|\psi\|=1}\left[\max_{b\in\mathcal{B}(\alpha)} \psi^T x - \min_{b\in\mathcal{B}(\alpha)} \psi^T x\right] = \sqrt{\max \text{eig}\ A^+ W^{-1} A^{+^T}} \tag{5.75}$$

$$= 2\alpha\sqrt{\lambda_{\text{max}}} \tag{5.76}$$

This relation shows how the measurement uncertainty, expressed by α, propagates to the estimated spectrum, x. The measurement uncertainty is amplified or compressed, depending on the value of λ_{max}. We can calculate λ_{max} based on prior information, so the sensitivity of the estimate can be anticipated before measurement.

Employing eq.(5.71), the condition for no-failure of the least-squares estimate becomes:

$$2\alpha\sqrt{\lambda_{\text{max}}} < p_{\text{cr}} \tag{5.77}$$

Expressing this as a reliability, the limiting value of the uncertainty parameter is:

$$\widehat{\alpha} = \frac{p_{\text{cr}}}{2\sqrt{\lambda_{\text{max}}}} \tag{5.78}$$

A large value of λ_{max}, which amplifies the measurement uncertainty, implies low reliability. Only small measurement uncertainty can be reliably tolerated when λ_{max} is large.

5.3.3 Estimating a Spectral Centroid

A shift in the input spectrum x defined in eq.(5.62), either to lower or higher frequencies, is indicative of a fault. Spectral shifts can often be detected by estimating the centroid of the spectrum. Our results so far are immediately applicable to evaluating the reliability of estimating the centroid of the input spectrum.

The centroid of the spectral vector x, which has K elements, is:

$$c = \frac{1}{K}\sum_{k=1}^{K} x_k \tag{5.79}$$

This can be represented as a projection of x. Define a vector $\mathbf{1} \in \Re^K$ whose elements are all ones. Then:

$$c = \frac{1}{K}\mathbf{1}^T x \tag{5.80}$$

The variation of the estimated centroid is evaluated from eq.(5.74) by choosing $\psi = \frac{1}{K}\mathbf{1}$, resulting in:

$$\operatorname*{opt}_{b\in\mathcal{B}(\alpha)} \frac{1}{K}\mathbf{1}^T x = \frac{1}{K}\mathbf{1}^T A^+\bar{b} \pm \frac{\alpha}{K}\sqrt{\mathbf{1}^T A^+ W^{-1} A^{+^T}\mathbf{1}} \tag{5.81}$$

Fractional Variation. The fractional variation of the centroid estimate can be as large as:

$$f = \frac{2\alpha\sqrt{\mathbf{1}^T A^+ W^{-1} A^{+^T}\mathbf{1}}}{\mathbf{1}^T A^+\bar{b}} \tag{5.82}$$

If we require that this fractional variation be no larger than a critical value, f_{cr}, the reliability of the estimate is:

$$\widehat{\alpha} = f_{\text{cr}}\frac{\mathbf{1}^T A^+\bar{b}}{2\alpha\sqrt{\mathbf{1}^T A^+ W^{-1} A^{+^T}\mathbf{1}}} \tag{5.83}$$

Hypothesis Testing. One may wish to detect specific spectral shifts with known associated centroid changes. This can be formulated as an hypothesis test. Consider N_h evenly spaced hypotheses for the value of the centroid. The ith hypothesis is that the centroid equals c_i:

$$H_i : \quad \frac{1}{K}\mathbf{1}^T A^+\bar{b} = c_i \tag{5.84}$$

The hypotheses are arranged in increasing order, $c_1 < \cdots < c_{N_h}$, and the increment between hypotheses is constant:

$$\Delta_c = \frac{c_{N_h} - c_1}{N_h - 1} \tag{5.85}$$

Given a measured centroid $c = \frac{1}{K}\mathbf{1}^T A^+ b$, we choose the closest hypothesis. That is, choose hypothesis H_k if:

$$|c - c_k| = \min_i |c - c_i| \tag{5.86}$$

If hypothesis H_i holds then the correct centroid is $\frac{1}{K}\mathbf{1}^T A^+ \overline{b} = c_i$, but the measurement can deviate from this by as much as:

$$\pm \frac{\alpha}{K}\sqrt{\mathbf{1}^T A^+ W^{-1} A^{+^T} \mathbf{1}} \tag{5.87}$$

due to uncertainty in b. The multi-hypothesis decision will err if the measurement deviates from the correct value by more than $\Delta_c/2$. The reliability of the diagnosis is the greatest uncertainty consistent with no-failure of the decision. This is obtained by equating the greatest deviation of the measurement to the greatest acceptable error, and solving for α:

$$\frac{\alpha}{K}\sqrt{\mathbf{1}^T A^+ W^{-1} A^{+^T} \mathbf{1}} = \frac{\Delta_c}{2} \quad \Longrightarrow \quad \widehat{\alpha} = \frac{K(c_{N_h} - c_1)}{2(N_h - 1)\sqrt{\mathbf{1}^T A^+ W^{-1} A^{+^T} \mathbf{1}}} \tag{5.88}$$

This relation shows that, for a given range of centroid values, $c_{N_h} - c_1$, the reliability decreases inversely with the number of hypotheses which are tested. There is a trade-off between reliability and resolution: as the number of hypotheses which are tested increases, the amount of tolerable failure-free uncertainty is reduced.

5.3.4 Reliability of "Regularized" Solutions

The starting point for least-squares analysis is the set of linear algebraic equations $Ax = b$. The least-squares approach provides a unique "best" solution, even when the equations are either conflicting or under-determined, by seeking an x which minimizes the quadratic form:

$$J = ||Ax - b||^2 \tag{5.89}$$

It often happens that the solution to this least-squares optimization is very sensitive to noise: small fluctuations in A or b can result in large variation in x. However, the solution can be stabilized or "regularized" by employing prior information to limit the variation of x.

For instance, suppose that prior information indicates that the solution should be something like $\overline{x}$. In this case, one can try to bias the solution in favor of $\overline{x}$, while still exploiting the information in the relation $Ax = b$. One way is to search for an x which minimizes:

$$J_r = ||Ax - b||^2 + \gamma \, ||x - \overline{x}||^2 \tag{5.90}$$

where γ is a positive constant which weights the preference for $x = \overline{x}$ against the preference for $Ax = b$. The choice of γ determines how strongly the prior information "pulls" the solution away from the nominal least-squares case.

There are various considerations for choosing the weighting term γ [70, 71]. In general, one purpose of the right-most term in eq.(5.90), the "regularizer", is to enhance the stability of the solution. We can express this in the spirit of robust reliability, and ask for the value of γ which results in the most reliable solution: the solution whose variation due to uncertainty in b is minimized.

Optimal solutions are found from the condition:

$$0 = \frac{\partial J_r}{\partial x} \tag{5.91}$$

which leads to the relation:

$$\left[A^T A - \gamma I\right] x = A^T b - \gamma \overline{x} \tag{5.92}$$

$A^T A$ is a real symmetric matrix, so it is diagonalized by an orthogonal matrix, V:

$$A^T A = V \Lambda V^T, \quad V V^T = V^T V = I \tag{5.93}$$

where Λ is diagonal. For simplicity we will assume that $A^T A$ is positive definite and hence that all the diagonal elements are positive: $0 < \lambda_1 \leq \cdots \leq \lambda_N$. Employing eq.(5.93), one can re-arrange eq.(5.92) as:

$$\left[\Lambda - \gamma I\right] V^T x = V^T A^T b - \gamma V^T \overline{x} \tag{5.94}$$

We will assume that the weighting term is not an eigenvalue:

$$\gamma \neq \lambda_i, \quad i = 1, \ldots, N \tag{5.95}$$

So the solution of (5.94) is:

$$x = V \left[\Lambda^{-1} - \frac{1}{\gamma} I\right] \left[V^T A^T b - \gamma V^T \overline{x}\right] \tag{5.96}$$

$$= \underbrace{\left[V \Lambda^{-1} V^T - \frac{1}{\gamma} I\right] A^T}_{C} \, b - \left[\gamma V \Lambda^{-1} V^T - I\right] \overline{x} \tag{5.97}$$

which defines the matrix C.

Now, suppose the uncertainty in b is represented by the spherical set:

$$\mathcal{B}(\alpha) = \left\{ b : \left(b - \overline{b}\right)^T \left(b - \overline{b}\right) \leq \alpha^2 \right\} \tag{5.98}$$

The maximum variation of projections of x along any direction ψ is expressed by eq.(5.76), where $\lambda_{\max}$ is the greatest eigenvalue of the matrix CC^T.

Using the relation:

$$(A^T A)^{-1} = V \Lambda^{-1} V^T \tag{5.99}$$

one can obtain the following expression for CC^T:

$$CC^T = \left[V \Lambda^{-1} V^T - \frac{1}{\gamma} I\right] A^T A \left[V \Lambda^{-1} V^T - \frac{1}{\gamma} I\right] \tag{5.100}$$

$$= V \left[\Lambda^{-1} - \frac{2}{\gamma} I + \frac{1}{\gamma^2} \Lambda\right] V^T \tag{5.101}$$

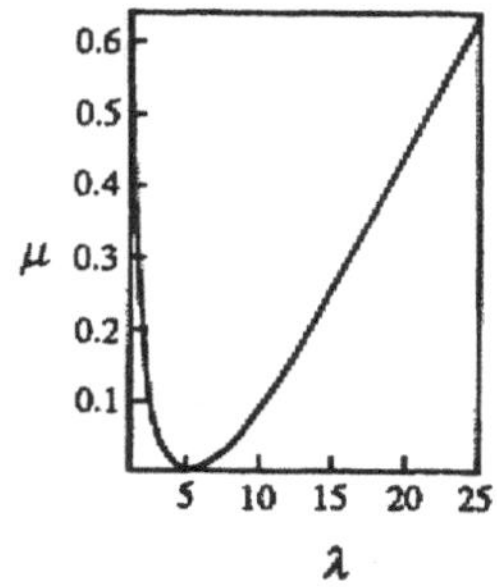

Figure 5.13: μ versus λ for $\gamma = 5$.

This shows that the eigenvalues of CC^T are precisely the diagonal elements of the diagonal matrix in square brackets in this expression, which we denote:

$$\mu_i(\gamma) = \frac{1}{\lambda_i} - \frac{2}{\gamma} + \frac{\lambda_i}{\gamma^2} = \frac{(\gamma - \lambda_i)^2}{\gamma^2 \lambda_i}, \quad i = 1, \ldots, N \tag{5.102}$$

So, for fixed weighting factor γ, the greatest variation of the projected solution for any projection direction is:

$$\max_{\|\psi\|=1} \left[\max_{b \in \mathcal{B}(\alpha)} \psi^T x - \min_{b \in \mathcal{B}(\alpha)} \psi^T x \right] = 2\alpha \sqrt{\max \text{ eig } CC^T} \tag{5.103}$$

$$= 2\alpha \sqrt{\max_i \mu_i} \tag{5.104}$$

To optimize the reliability of the regularized least-squares solution, we must choose γ to minimize the maximum eigenvalue. In other words, we must solve the following optimization problem:

$$\min_\gamma \max_i \mu_i(\gamma) \tag{5.105}$$

The partial derivative of μ with respect to λ is:

$$\frac{\partial \mu}{\partial \lambda} = \frac{\lambda^2 - \gamma^2}{\lambda^2 \gamma^2} \tag{5.106}$$

In other words, for fixed γ, $\mu_i(\gamma)$ increases with the difference between γ and λ_i, as shown in fig. 5.13. Consequently, the inner maximum in (5.105) occurs at either the least or the greatest eigenvalue:

$$\max_i \mu_i(\gamma) = \max \left\{ \frac{(\gamma - \lambda_1)^2}{\gamma^2 \lambda_1}, \frac{(\gamma - \lambda_N)^2}{\gamma^2 \lambda_N} \right\} \tag{5.107}$$

This furthermore implies that the value of γ which minimizes the maximum will lie between λ_1 and λ_N. As γ increases, μ_1 increases and μ_N decreases, while if γ decreases the reverse situation holds. In other words, the value of γ which minimizes the maximum of $\{\mu_1, \mu_2\}$ is the value of γ at which equality holds:

$$\mu_1 = \mu_N \quad \text{or} \quad \frac{(\gamma - \lambda_1)^2}{\gamma^2 \lambda_1} = \frac{(\gamma - \lambda_N)^2}{\gamma^2 \lambda_N} \tag{5.108}$$

Solving for γ and choosing the positive root, one finds that the optimal weighting factor, which makes the regularized least-squares solution least sensitive to uncertainty in b, is the geometric mean of the least and greatest eigenvalues of $A^T A$:

$$\widehat{\gamma} = \sqrt{\lambda_1 \lambda_N} \tag{5.109}$$

The greatest eigenvalue of CC^T for this choice of the regularization weighting factor is:

$$\max_i \mu_i = \mu_1(\widehat{\gamma}) = \mu_N(\widehat{\gamma}) \tag{5.110}$$

$$= \left(\frac{1}{\sqrt{\lambda_1}} - \frac{1}{\sqrt{\lambda_N}} \right)^2 \tag{5.111}$$

Combining this with relation (5.104) leads finally to an expression for the greatest variation of the *regularized* projected solution, for any projection direction:

$$\max_{\|\psi\|=1} \left[\max_{b \in \mathcal{B}(\alpha)} \psi^T x - \min_{b \in \mathcal{B}(\alpha)} \psi^T x \right] = 2\alpha \left(\frac{1}{\sqrt{\lambda_1}} - \frac{1}{\sqrt{\lambda_N}} \right) \tag{5.112}$$

Our development, from eq.(5.96) on, has assumed that $\gamma > 0$. The ordinary least-squares solution arises when $\gamma = 0$, for which we must return to eq.(5.92). Now instead of eq.(5.97) we have:

$$x = \underbrace{(A^T A)^{-1} A^T}_{C} \ b \tag{5.113}$$

so CC^T now becomes:

$$CC^T = (A^T A)^{-1} \tag{5.114}$$

whose eigenvalues are $1/\lambda_N \le \cdots \le 1/\lambda_1$. So the greatest variation of the *unregularized* projected solution becomes, as in eq.(5.103):

$$\max_{\|\psi\|=1} \left[\max_{b \in \mathcal{B}(\alpha)} \psi^T x - \min_{b \in \mathcal{B}(\alpha)} \psi^T x \right] = 2\alpha \sqrt{\frac{1}{\lambda_1}} \tag{5.115}$$

The ratio of the maximum regularized projection, eq.(5.112), to the unregularized case, eq.(5.115), gives some indication of the best enhancement of the stability achievable by this procedure:

$$\rho = \sqrt{\frac{\mu_1(\widehat{\gamma})}{1/\lambda_1}} \tag{5.116}$$

$$= 1 - \sqrt{\frac{\lambda_1}{\lambda_N}} \tag{5.117}$$

This varies from zero to unity, depending on the ratio of the least to the greatest eigenvalues of $A^T A$. We note that, when $A^T A$ is very ill-conditioned, so that $\lambda_1/\lambda_N \ll 1$, the regularization provides little reduction in the range of variation of the solution, if the uncertainty in b is represented by the spherical uncertainty-model of eq.(5.98).

5.4 Multi-Hypothesis Diagnosis of a Crack

Measurement of eigenvalues and mode shapes can be used in many ways to detect and diagnose cracks in elastically vibrating structures. In this section we will evaluate the robust reliability of a multi-hypothesis algorithm for diagnosing small cracks which is based on modal measurements.

The basic structure of the diagnosis is again the estimation of the vector x from a relation of the form $Ax = b$, where uncertainties are associated with both A and b. The multi-hypothesis algorithm however is very different from the least-squares approach discussed in section 5.3.

Cracks usually open and close during each cycle of vibration, resulting in quite complicated time-varying dynamical effects. However, much experimental and theoretical work has shown that quite often a small crack can be represented fairly well as though it produced a constant reduction in stiffness at the site of the crack. We will adopt this approach here.

5.4.1 The Eigenvalue Equation

The eigenvalue equation for mechanical vibration relates the mass and stiffness matrices, M and K, via an eigenvector (or mode-shape vector) y and an eigenvalue λ (which is the square of a natural frequency):

$$(\lambda M + K)y = 0 \tag{5.118}$$

M and K are real, symmetric, positive definite[3] $N \times N$ matrices, where N is the number of degrees of freedom of the dynamical model. N is often quite large, in which case it is not feasible to measure all the modes. We will assume that L modes are measured, and that the mode shapes are denoted by the columns of a matrix $Y = [y_1, \ldots, y_L]$ and the corresponding eigenvalues

[3] M will be positive semi-definite if rigid body degrees of freedom are included in the equations of motion.

are stored in the diagonal matrix $\Lambda = \text{diag}(\lambda_1, \ldots, \lambda_L)$. Eq.(5.118) can be expressed for all L measured modes as:

$$(M\Lambda + K)Y = 0 \tag{5.119}$$

The mass matrix M is known at least approximately and the eigenvalues Λ and mode shapes Y are measured, while the stiffness matrix K must be up-dated to locate and diagnose any cracks which may be present. In other words this relation is of the form $Ax = b$ where A and b, based on M, Λ and Y, are measured but uncertain, and x, based on K, must be estimated.

5.4.2 The Multi-Hypothesis Algorithm

Direct inversion of eq.(5.119) to find the stiffness matrix K tends to be unstable to noise in the measured mode shapes, Y. (We will discuss the source of this ill-conditioning in section 5.6.) Rather than employing a least-squares method for inverting this relation, we will seek to identify the stiffness changes by testing a range of hypothesized stiffness matrices.

The crack-free stiffness matrix is K_0, and we postulate a collection of N_h hypothesized cracks, each represented by a modification K_h to the nominal stiffness matrix:

$$K = K_0 + K_h, \quad h = 1, \ldots, N_h \tag{5.120}$$

For local stiffness changes these hypothesized matrices can often be quite sparse and simple. A typical form for the hypothesized modifications is:

$$K_h = k_h u_h u_h^T \tag{5.121}$$

where k_h is a scalar stiffness parameter and u_h is a vector which indicates the location of the crack. For simple stiffness modifications such as these, a reasonably comprehensive list of hypotheses is not too long and can be tested by the multi-hypothesis method. In light of eq.(5.119), the matrix $(M\Lambda + K_0 + K_n)Y$ will be nearly zero if the nth hypothesis is nearly correct, and large otherwise. The multi-hypothesis algorithm for identifying the stiffness matrix is to adopt the nth hypothesis if:

$$||(M\Lambda + K_0 + K_n)Y|| = \min_h ||(M\Lambda + K_0 + K_h)Y|| \tag{5.122}$$

where $||A||$ is the euclidean norm of the matrix A, which is the sum of the euclidean norms of the columns of $A = [a_1, \ldots, a_L]$:

$$||A||^2 = \sum_{i=1}^{L} a_i^T a_i \tag{5.123}$$

For notational convenience let us define the nominal failure-free matrix from the lefthand side of eq.(5.119):

$$J_0 = (M\Lambda + K_0)Y \tag{5.124}$$

whose columns we denote as $j_{0,1}, \ldots, j_{0,L}$. Similarly, define the hth hypothesized modification of J_0 as:

$$J_h = J_0 + K_h Y \tag{5.125}$$

whose columns are $j_{0,i} + K_h y_i$ for $i = 1, \ldots, L$. The multi-hypothesis algorithm, eq.(5.122), can be expressed as follows. Choose the nth hypothesis if:

$$\begin{aligned} \|J_n\|^2 &= \min_h \|J_h\|^2 && (5.126) \\ &= \min_h \sum_{i=1}^{L} \|j_{0,i} + K_h y_i\|^2 && (5.127) \end{aligned}$$

5.4.3 Performance Criterion for the Diagnosis

Eigenvalues can, in some circumstances, be measured quite accurately. Furthermore prior information can be brought to bear in formulating an uncertainty model for the mode shapes. We will consider an envelope-bound convex model for the mode-shape uncertainty, in which the fractional variation of each mode shape is bounded by the uncertainty parameter α:

$$\mathcal{U}(\alpha) = \left\{ Y : \left| y_{ij} - \overline{y}_{ij} \right| \le \alpha \left| \overline{y}_{ij} \right|, \quad i = 1, \ldots, L, \quad j = 1, \ldots, N \right\} \tag{5.128}$$

where $\overline{y}_i$ is the anticipated form of the ith mode shape. Let $\overline{Y} = [\overline{y}_1, \ldots, \overline{y}_L]$. Then we can write $Y = Z + \overline{Y}$ where Z is the deviation of Y from the anticipated modal matrix $\overline{Y}$. Now $\mathcal{U}(\alpha)$ can be written:

$$\mathcal{U}(\alpha) = \left\{ Y = Z + \overline{Y} : \left| z_{ij} \right| \le \alpha \left| \overline{y}_{ij} \right|, \quad i = 1, \ldots, L, \quad j = 1, \ldots, N \right\} \tag{5.129}$$

The capability of the multi-hypothesis algorithm to distinguish between hypothesized cracks is limited by the variation of J_h due to mode-shape uncertainty. The greater the variation of J_h, the fewer is the number of hypotheses which can be unequivocally distinguished. This suggests a criterion for evaluating the reliability of the diagnostic algorithm.

The matrix J_h can be written:

$$J_h = J_0 + K_h \overline{Y} + K_h Z \tag{5.130}$$

where the variation of J_h due to uncertainty is concentrated in the term $K_h Z$, where Z varies around the zero matrix. For a given set of hypothesized stiffness matrices, $K_1, \ldots, K_{N_h}$, let us suppose that the norms $\|J_0 + K_1 \overline{Y}\|, \ldots, \|J_0 + K_{N_h} \overline{Y}\|$ are spaced so that the minimum difference between them is Δ_{cr}. The multi-hypothesis algorithm will tend to perform satisfactorily if $\|K_h Z\|$ varies by no more than about $\Delta_{\text{cr}}/2$ for $h = 1, \ldots, N_h$. In other words, we will adopt the following condition for failure of the diagnostic algorithm:

$$\max_{Z\in\mathcal{U}(\alpha)} \|K_h Z\| - \min_{Z\in\mathcal{U}(\alpha)} \|K_h Z\| > \frac{\Delta_{\rm cr}}{2}, \quad \text{for some } h = 1, \ldots, N_h \tag{5.131}$$

The reliability of the diagnostic algorithm is the greatest value of the uncertainty parameter, α, consistent with no-failure. This is not the only possible criterion, but it is both plausible and tractable.

5.4.4 A Useful Theorem

We will now present a theorem which will be useful in evaluating the reliability of the multi-hypothesis diagnosis. We must precede the theorem with a simple lemma, based on the Cauchy inequality, which is discussed in example 2 of chapter 2 (p.18).

Lemma 1 *For any vector v, the euclidean norm of v can be expressed as an extremal projection along a unit vector. That is:*

$$\max_{\|\psi\|=1} \psi^T v = \|v\| \tag{5.132}$$

where $\|v\| = \sqrt{v^T v}$.

Proof. The Cauchy inequality states that $\psi^T v$ is bounded as:

$$(\psi^T v)^2 \le \|\psi\|^2 \, \|v\|^2 \tag{5.133}$$

with equality if ψ is parallel to v. The vector ψ is constrained to the unit sphere and thus can lie in any direction. Hence ψ can be chosen parallel to v. Consequently, equality holds in (5.133) for this choice of ψ, whose norm equals unity. ∎

Theorem 1 *Let $f(u)$ be a continuous vector function defined on a closed and bounded set $\mathcal{U}$. The maximal value of the euclidean norm of f on $\mathcal{U}$ can be expressed as an extremal projection along a unit vector:*

$$\max_{\|\psi\|=1} \max_{u\in\mathcal{U}} \psi^T f(u) = \max_{u\in\mathcal{U}} \|f(u)\| \tag{5.134}$$

Proof. Since f is continuous and $\mathcal{U}$ is closed and bounded, $f(u)$ is bounded on $\mathcal{U}$ and, for any vector ψ, the function $\psi^T f(u)$ has a minimum and a maximum value on $\mathcal{U}$.

Let M denote the double maximum on the lefthand side of (5.134). Suppose there is an element $\widetilde{u} \in \mathcal{U}$ such that:

$$\|f(\widetilde{u})\| > M \tag{5.135}$$

Then, by lemma 1:

$$\max_{||\psi||=1} \psi^T f(\widetilde{u}) = ||f(\widetilde{u})|| \tag{5.136}$$

Consequently,

$$\max_{||\psi||=1} \psi^T f(\widetilde{u}) > M \tag{5.137}$$

which is a contradiction. Hence supposition (5.135) is false and so:

$$\max_{u \in \mathcal{U}} ||f(u)|| \leq M \tag{5.138}$$

for all $u \in \mathcal{U}$. We can immediately see that this cannot be a strict inequality. $\mathcal{U}$ is closed and bounded so the maxima in M exist. In other words, there is a $\widehat{u} \in \mathcal{U}$ and a $||\widehat{\psi}|| = 1$ such that:

$$M = \widehat{\psi}^T f(\widehat{u}) \tag{5.139}$$

But, employing lemma 1:

$$\widehat{\psi}^T f(\widehat{u}) \leq \max_{||\psi||=1} \psi^T f(\widehat{u}) = ||f(\widehat{u})|| \tag{5.140}$$

Hence:

$$M \leq \max_{u \in \mathcal{U}} ||f(u)|| \tag{5.141}$$

Relations (5.138) and (5.141) together prove eq.(5.134). ■

Example 3 Let us consider an example of theorem 1. Let $f(u) = u$ and let $\mathcal{U}$ we an ellipsoidal set of vectors:

$$\mathcal{U} = \{u : \ u^T W u \leq \alpha^2\} \tag{5.142}$$

where W is a positive definite, real, symmetric matrix. We will find the maximum norm of $f(u)$ on $\mathcal{U}$ first by using the method of Lagrange multipliers and then by employing theorem 1.

(1) *Using Lagrange optimization.* To maximize $||f(u)||$ on $\mathcal{U}$ we must solve the following optimization problem:

$$\max u^T u \quad \text{subject to} \quad u^T W u \leq \alpha^2 \tag{5.143}$$

The extrema occur on the boundary. Adjoining the constraint to form $J = u^T u + \lambda(\alpha^2 - u^T W u)$, the condition for an extremum is:

$$0 = \frac{\partial J}{\partial u} \tag{5.144}$$

which implies:

$$(I - \lambda W)u = 0 \tag{5.145}$$

In other words, the optimal solutions occur when u is an eigenvector of W and when $1/\lambda$ is the corresponding eigenvalue. Denote the eigenvalues of W by $\mu_1 \leq \cdots \leq \mu_N$. They are all positive since W is positive definite. Employing the constraint one obtains the final result:

$$\max_{u \in \mathcal{U}} \|f(u)\| = \frac{\alpha}{\sqrt{\mu_1}} \tag{5.146}$$

(2) *Using theorem 1.* We must evaluate the double maximum:

$$\max_{\|\psi\|=1} \max_{u \in \mathcal{U}} \psi^T f(u) \tag{5.147}$$

Consider first the inner maximum, for fixed ψ:

$$\max \psi^T u \quad \text{subject to} \quad u^T W u \leq \alpha^2 \tag{5.148}$$

This is the optimization of a linear function on $\mathcal{U}$, unlike (5.143) which optimizes a quadratic function.[4] The maximum again occurs on the boundary. Adjoining the constraint, which is now an equality, define: $J = \psi^T u + \lambda(\alpha^2 - u^T W u)$. Eq.(5.144) is again the condition for an extremum, which now implies:

$$0 = \psi - 2\lambda W u \tag{5.149}$$

Employing the constraint one finds the maximum projection along ψ to be:

$$\max_{u \in \mathcal{U}} \psi^T f(u) = \alpha \sqrt{\psi^T W^{-1} \psi} \tag{5.150}$$

We now proceed to the outer maximization. Since ψ is a unit vector, the radical in eq.(5.150) varies over the field of values [56] of W^{-1}, that is, between the least and greatest eigenvalues of W^{-1}, which are $1/\mu_N$ and $1/\mu_1$, respectively. So the outer maximum occurs at the greatest eigenvalue of W^{-1}:

$$\max_{\|\psi\|=1} \max_{u \in \mathcal{U}} \psi^T f(u) = \frac{\alpha}{\sqrt{\mu_1}} \tag{5.151}$$

just as in eq.(5.146). ∎

5.4.5 Reliability of the Diagnosis

We are now in a position to evaluate the reliability of the multi-hypothesis diagnosis algorithm, eq.(5.122) or eq.(5.126). The algorithm is considered to fail if inequality (5.131) holds for some value of the hypothesis-index h. Let V represent the righthand side of this relation: the maximum variation of the norm of the uncertain term, $K_h Z$. This variation depends on the uncertainty parameter, α, and the robust reliability of the algorithm is the greatest value

[4] It is sometimes easier to optimize a linear function than a quadratic one, as we will see in our subsequent example in section 5.4.5.

of α consistent with no failure. We evaluate the reliability by equating $V(\alpha)$ to the critical value, $\Delta_{cr}/2$, and solving for α. In other words, the reliability $\widehat{\alpha}$ is the solution of:

$$V(\widehat{\alpha}) = \frac{\Delta_{cr}}{2} \tag{5.152}$$

Theorem 1 will be useful in evaluating $V(\alpha)$, which we now proceed to do.

We must determine the least and greatest values of the norm $||K_h Z||$, which depends on the uncertain quantity Z, whose variation is expressed by the convex model of eq.(5.129). The minimum is clearly zero since Z varies around the zero matrix. Now for the maximum.

$K_h Z$ is an $N \times L$ matrix, but we can concatenate its columns into a single long vector, which we denote as follows. The ith column of Z is z_i, so the ith column of $K_h Z$ is $K_h z_i$. Consequently the vector form of $K_h Z$ becomes:

$$\mathrm{vec}(K_h Z) = \left[(K_h z_1)^T, \ldots, (K_h z_L)^T\right]^T \tag{5.153}$$

The correspondence between our optimization problem and theorem 1 is obtained by replacing the uncertain quantity u by Z, and the vector function $f(u)$ by $\mathrm{vec}(K_h Z)$:

$$u \Longrightarrow Z, \qquad f(u) \Longrightarrow \mathrm{vec}(K_h Z) \tag{5.154}$$

The euclidean norms of the vector $\mathrm{vec}(K_h Z)$ and of the matrix $K_h Z$ are identical, as seen from eq.(5.123).

Theorem 1 states that the maximum norm of $\mathrm{vec}(K_h Z)$ can be evaluated as:

$$\max_{Z \in \mathcal{U}(\alpha)} ||K_h Z|| = \max_{||\psi||=1} \max_{Z \in \mathcal{U}(\alpha)} \psi^T \mathrm{vec}(K_h Z) \tag{5.155}$$

We begin with the inner optimization, for which we must find the extremal values of $\psi^T \mathrm{vec}(K_h Z)$. Now ψ is a vector with NL elements, but we can break it into L blocks, $\psi^T = (\psi_1^T, \ldots, \psi_L^T)$, and express $\psi^T \mathrm{vec}(K_h Z)$ as:

$$\psi^T \mathrm{vec}(K_h Z) = \sum_{i=1}^{L} \psi_i^T K_h z_i \tag{5.156}$$

We must ultimately satisfy the requirement $||\psi|| = 1$, but other than that, the blocks $\psi_1, \ldots, \psi_L$ can be chosen independently.

We can write the inner product explicitly as:

$$\psi^T \mathrm{vec}(K_h Z) = \sum_{i=1}^{L} \sum_{j=1}^{N} \psi_i^T (K_h)_j z_{ij} \tag{5.157}$$

where $(K_h)_j$ is the jth column of K_h. The convex model allows each element z_{ij} to vary within an interval:

$$-\alpha|\overline{y}_{ij}| \leq z_{ij} \leq \alpha|\overline{y}_{ij}| \tag{5.158}$$

The maximum value of $\psi^T \text{vec}(K_h Z)$ is obtained by choosing z_{ij} at its upper or lower limit, depending on whether $\psi_i^T (K_h)_j$ is positive or negative, respectively:

$$\max_{Z \in \mathcal{U}(\alpha)} \psi^T \text{vec}(K_h Z) = \sum_{i=1}^{L} \sum_{j=1}^{N} \psi_i^T (K_h)_j \text{sgn}\left[\psi_i^T (K_h)_j\right] \alpha |\overline{y}_{ij}| \tag{5.159}$$

$$= \alpha \sum_{i=1}^{L} \sum_{j=1}^{N} |\psi_i^T (K_h)_j| \cdot |\overline{y}_{ij}| \tag{5.160}$$

where $\text{sgn}(x)$ equals the sign of x.

To find the maximum norm of $K_h Z$ we must maximize (5.160) for $||\psi|| = 1$. This requires some numerical effort, but the result is independent of the uncertainty parameter, α. Let S denote the maximum value of the double sum in (5.160), so that the maximum norm is:

$$\max_{Z \in \mathcal{U}(\alpha)} ||K_h Z|| = \alpha S \tag{5.161}$$

Since the minimum norm is zero, this is precisely $V(\alpha)$. Hence, employing eq.(5.152), the reliability is:

$$\widehat{\alpha} = \frac{\Delta_{\text{cr}}}{2S} \tag{5.162}$$

5.5 Robust Reliability of Model-Order Determination

Determination of the order or dimension of a model is a particularly difficult and noise-sensitive system identification problem.[5] Many procedures, such as Akaike's final prediction error method, are discussed in the literature [25, 63, 87]. In this section we will outline an application of robust reliability analysis to the problem of choosing the order of a polynomial for fitting uncertain measured data points.

5.5.1 Formulation

In this subsection we formulate the basic reliability analysis for determining the model order, and in the next subsection we examine two examples.

We have N measured but uncertain data points, $(x_1, y_1), \ldots, (x_N, y_N)$, which we will fit with a polynomial of order M:

$$f(x) = \sum_{m=0}^{M} a_m x^m \tag{5.163}$$

[5] I am indebted to Prof. S. Braun of the Technion for directing my attention to this problem.

We will choose the polynomial coefficients $a_0, \ldots, a_M$ to minimize the squared error of the fit:

$$S_M = \min_{a_0, \ldots, a_M} \sum_{n=1}^{N} [y_n - f(x_n)]^2 \qquad (5.164)$$

S_M, the least-squared error, is a measure of the quality of the polynomial fit of order M.

We wish to choose the order M of the fitting polynomial. If the number of coefficients, $M+1$, equals the number of data points, N, then the least-squared error S_M will vanish. But it is often impractical and unnecessary to choose such a high-order polynomial. We will develop a robust measure of the reliability of the fit, which allows one to compare different model-orders.

Let S_{cr} denote the greatest acceptable value of the least squared error. That is, to use the language of reliability analysis, we say that the polynomial fit of order M has failed if:

$$S_M > S_{\text{cr}} \qquad (5.165)$$

The uncertainty of the measured data points is represented by a convex model, $\mathcal{U}(\alpha)$. In other words, the data points $(x_1, y_1), \ldots, (x_N, y_N)$ vary on the set $\mathcal{U}(\alpha)$. We obtain a value of the least-squared error S_M for any given set of data points in $\mathcal{U}(\alpha)$. Furthermore, we can evaluate the greatest magnitude of the least-squared error which could be obtained with any data in the convex model. Let us denote the maximum of the least-squared error by:

$$\widehat{S}_M(\alpha) = \max_{\mathcal{U}(\alpha)} S_M \qquad (5.166)$$

The robust reliability of model-order M is the greatest value of the uncertainty parameter α which is consistent with performance criterion (5.165). That is, equating $\widehat{S}_M(\alpha)$ to S_{cr} and solving for α yields the robust reliability of the Mth-order polynomial fit:

$$\widehat{S}_M(\alpha) = S_{\text{cr}} \quad \Longrightarrow \quad \widehat{\alpha}_M \qquad (5.167)$$

We will illustrate this analysis for two different choices of convex uncertainty models.

However, before proceeding to examples, let us make a few general observations. Examination of eq.(5.166) reveals that, for fixed moder-order M, $\widehat{S}(\alpha)$ will increase monotonically with α, as indicated schematically in fig. 5.14(a). This results from the property of nesting of convex models. One of the basic characteristics of all the convex models we have used is that as a convex model expands by increasing α, it contains all the smaller sets of the same family. That is, if $\alpha < \alpha'$ then $\mathcal{U}(\alpha)$ is a subset of $\mathcal{U}(\alpha')$:

$$\alpha < \alpha' \quad \Longrightarrow \quad \mathcal{U}(\alpha) \subset \mathcal{U}(\alpha') \qquad (5.168)$$

The second point to note is that one would usually expect $\widehat{S}_M(\alpha)$ to decrease with M, for fixed uncertainty α, as shown in fig. 5.14(b). For a

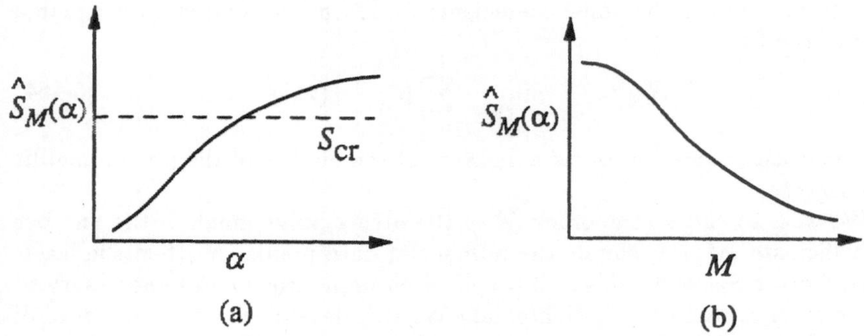

Figure 5.14: Schematic dependence of $\widehat{S}_M(\alpha)$ on α and on M.

given set of data, the least-squared error S_M from eq.(5.164) decreases with M:

$$S_{M+1} \leq S_M \tag{5.169}$$

This inequality holds because the Mth order polynomial is a special case of the $(M+1)$th order polynomial. However, this does not unequivocally imply that $\widehat{S}_{M+1}(\alpha)$ will be less than $\widehat{S}_M(\alpha)$, and when $M+1$ is near N (the number of data points) so that $\widehat{S}_{M+1}(\alpha)$ is nearly zero, we might observe the reverse inequality. Nonetheless, the general trend of $\widehat{S}_M$ versus M will tend to be as shown in fig. 5.14(b).

From fig. 5.14(a) we see how the robust reliability of an Mth order polynomial fit is evaluated as the largest value of α consistent with no-failure according to condition (5.165). The straight line $\widehat{S}_M = S_{\rm cr}$ intersects the curve $\widehat{S}_M(\alpha)$ at a single point, whose abscissa is the robust reliability.

Combining figs. 5.14(a) and (b) as in fig. 5.15(a) shows $\widehat{S}_M(\alpha)$ versus α for a range of model-orders M. The value of M increases from the top to the bottom curve, in accordance with the discussion of fig. 5.14(b). Each of these curves intersects the critical value $S_{\rm cr}$ at a single point, whose abscissa is the robust reliability, $\widehat{\alpha}_M$, for polynomial fit of that model-order. From this figure we see that $\widehat{\alpha}_M$ increases monotonically with model-order, as shown schematically in fig. 5.15(b).

5.5.2 Examples

We first formulate a convenient expression for the least-squared error of the polynomial fit.

S_M can be expressed very succinctly with the help of a few definitions. The measured data are the N couplets $(x_1, y_1), \ldots, (x_N, y_N)$. Let $\vec{y}$ denote the vector of y-values:

$$\vec{y} = (y_1, \ldots, y_N)^T \tag{5.170}$$

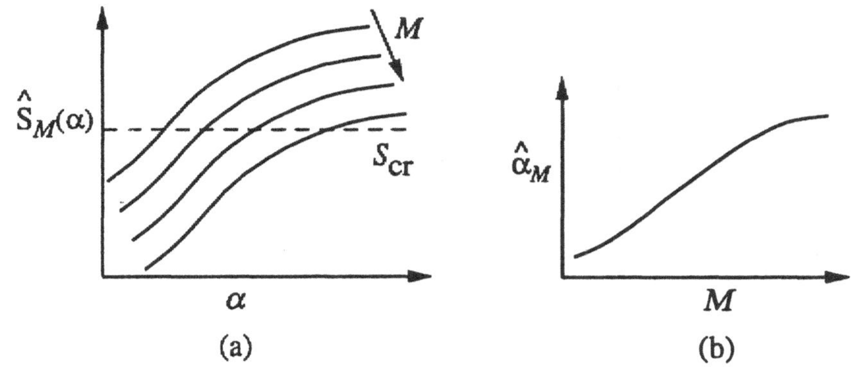

Figure 5.15: (a) Schematic dependence of $\widehat{S}_M(\alpha)$ on α for various values of the model-order M. (b) Variation of $\widehat{\alpha}_M$ with M.

Also, let $\vec{x}_i$ denote the vector of powers of the value x_i:

$$\vec{x}_i = \left(x_i^0, x_i^1, x_i^2, \ldots, x_i^M\right)^T, \quad i = 1, \ldots, N \tag{5.171}$$

Finally, define the following two matrices:

$$X = [\vec{x}_1, \ldots, \vec{x}_N] \in \Re^{(M+1)\times N} \tag{5.172}$$

$$C = \sum_{n=1}^{N} \vec{x}_n \vec{x}_n^T = XX^T \in \Re^{(M+1)\times(M+1)} \tag{5.173}$$

If $M + 1 \leq N$ and $x_i \neq x_j$ for all $i \neq j$, then X has full row rank because it is the first $M + 1$ rows of a non-singular Vandermonde matrix. Likewise C will be non-singular.

It is now an elementary matter to show that the least-squared error of the Mth-order polynomial fit can be expressed as:

$$S_M = \vec{y}^T \underbrace{\left[I - X^T C^{-1} X\right]}_{\Xi} \vec{y} \tag{5.174}$$

$$= \vec{y}^T \Xi \vec{y} \tag{5.175}$$

where the real symmetric matrix[6] Ξ is defined in eq.(5.174).

The maximum value of the least-squared error of the polynomial fit of order M is the solution of the following general optimization problem:

$$\widehat{S}_M(\alpha) = \max_{\mathcal{U}(\alpha)} \vec{y}^T \Xi \vec{y} \tag{5.176}$$

[6]We note that Ξ is a "projector matrix", which projects into the column null space of X. Consequently, as for all projector matrices, the eigenvalues of Ξ are 0's and 1's.

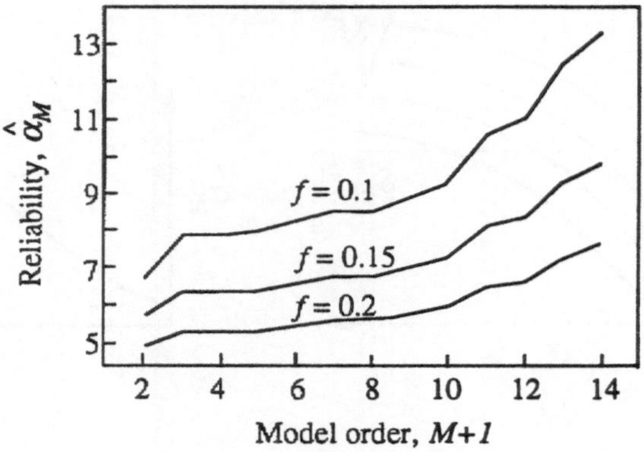

Figure 5.16: Reliability $\widehat{\alpha}_M$ versus number of fitting coefficients $M+1$. Number of data points is $N = 15$. $S_{cr} = 1$. $f = 0.1, 0.15, 0.2$.

We will now consider two examples based on different choices of the convex model $\mathcal{U}(\alpha)$.

Example 4 Suppose that the x-coordinate data are certain and that the measured vector $\vec{y}$ belongs to an ellipsoid centered at the origin:

$$\mathcal{U}(\alpha) = \left\{\vec{y} : \vec{y}^T W \vec{y} \leq \alpha^2\right\} \tag{5.177}$$

The maximum error $\widehat{S}_M(\alpha)$ is the solution of:

$$\max \vec{y}^T \Xi \vec{y} \quad \text{subject to} \quad \vec{y}^T W \vec{y} \leq \alpha^2 \tag{5.178}$$

We have encountered this optimization problem previously in section 4.5.2 and the solution is:

$$\widehat{S}_M(\alpha) = \alpha^2 \max \text{ eig } \left[W^{-1/2} \Xi W^{-1/2}\right] \tag{5.179}$$

Equating this result to the critical value of the error as in relation (5.167) and solving for the uncertainty parameter leads to the following expression for the robust reliability of the Mth-order polynomial fit:

$$\widehat{\alpha}_M = \sqrt{\frac{S_{cr}}{\max \text{ eig } \left[W^{-1/2} \Xi W^{-1/2}\right]}} \tag{5.180}$$

Fig. 5.16 shows the variation of $\widehat{\alpha}_M$ versus the order of the fitting function. The three curves correspond to different choices of the weighting matrix W, which determines the shape of the uncertainty-ellipsoid in the convex model of eq.(5.177). W is chosen to be proportional to a diagonal matrix as follows:

$$W = c \text{ diag}(w_1, \ldots, w_N), \quad w_n = 1 - \frac{n-1}{N-1}(1-f) \tag{5.181}$$

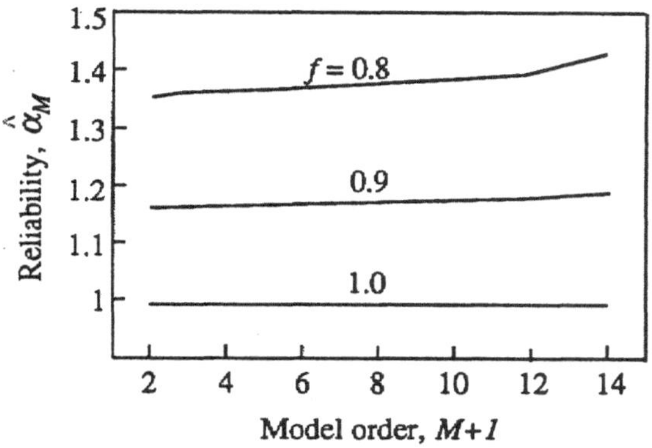

Figure 5.17: Reliability $\widehat{\alpha}_M$ versus number of fitting coefficients $M+1$. Number of data points is $N = 15$. $S_{\rm cr} = 1$. $f = 0.8, 0.9, 1.0$.

The value of f is different for each of the three convex models, and c is chosen so that the three ellipsoids have the same volume:

$$c = \left(\prod_{n=1}^{N} w_n \right)^{-1/2} \tag{5.182}$$

The eccentricity of the ellipsoid increases as f decreases from unity, while $f = 1$ generates a spheroid.

The robust reliability, $\widehat{\alpha}_M$, is the greatest value of the uncertainty parameter α in the convex model of eq.(5.177) which is consistent with no-failure, according to the criterion of eq.(5.165). In all cases the reliability of the fit increases as the number of polynomial coefficients approaches the number the data points, $N = 15$. (The case $M + 1 = N$ is not shown, for in this case Ξ vanishes, $\widehat{S}_M = 0$ and $\widehat{\alpha}_M = \infty$.) As the eccentricity grows, the reliability varies more strongly with model-order, so the curve for $f = 0.1$ varies most strongly with model order, while the curve $f = 0.2$ varies the least.

Furthermore, for fixed model order, the reliability increases with eccentricity of the uncertainty-ellipsoid. The eccentricity expresses prior information about how the uncertain data vectors $\vec{y}$ tend to cluster. A highly eccentric convex model implies a tendency for correlation between magnitude and direction of $\vec{y}$. At the other extreme, a completely spheroidal shape implies no knowledge about preferred directions of the data vector. The results of fig. 5.16 show that a highly eccentric convex model exploits a high-order model better than a low-order model. At the other extreme, spheroidal models show little or no improvement with model-order,[7] as seen in fig. 5.17. In addition, the values of $\widehat{\alpha}_M$ in fig. 5.17 are much less than in fig. 5.16.

[7]When the convex model is a sphere, then W is the identity matrix and the maximum eigenvalue in eq.(5.180) is unity regardless of the value of M, as commented in the footnote on p.137.

Consider again the case $f = 0.1$ in fig. 5.16. The highest-order model has a reliability index nearly twice as great as the lowest-order model. This means that the high-order model can tolerate more uncertainty without failing than the low-order model. However, this by itself does not indicate how large a value of the robust reliability is needed or desirable. We will return to consider the subjective interpretation of robust reliability in chapter 9. ∎

Example 5 Consider now an envelope-bound convex model. We suppose that the x-coordinate data are certain and that the measured y_i can deviate continuously on intervals of unknown size α. If $\overline{y}_i$ is the nominal or the measured value, then the envelope-bound convex model is the set of intervals:

$$\mathcal{U}(\alpha) = \{\vec{y}: \ |y_i - \overline{y}_i| \leq \alpha, \quad i = 1, \ldots, N\} \tag{5.183}$$

We do not need to know the range of uncertainty, α, and our analysis expresses the robust reliability of the polynomial fit as the greatest value of the uncertainty parameter α which is consistent with no-failure according to performance criterion (5.165).

In this example we have no convenient analytical solution for the greatest least-squared error, $\widehat{S}_M(\alpha)$:

$$\widehat{S}_M(\alpha) = \max \vec{y}^T \Xi \vec{y} \quad \text{subject to} \quad |y_i - \overline{y}_i| \leq \alpha, \ i = 1, \ldots, N \tag{5.184}$$

$\widehat{S}_M(\alpha)$ increases monotonically with the uncertainty parameter, α, so the optimization in (5.184) must be solved numerically for progressively increasing values of α until $\widehat{S}_M(\alpha)$ reaches the critical value, S_{cr}. The corresponding value of α is the robust reliability for Mth-order polynomial fit with interval uncertainty of unknown magnitude in the measurements. ∎

5.6 Ill-Posed Problems

In the previous sections we have encountered fault-diagnosis and system-identification problems which require the solution for a vector x of linear equations of the form:

$$Ax = y \tag{5.185}$$

where A and/or y are based on measurements.

Let us consider in abstract terms the difficulties one may confront in solving eq.(5.185). If there were no noise in the measurements then an exact solution to eq.(5.185) will exist if eq.(5.185) accurately represents a physical system. However, because both A and y are noisy it is possible that a solution to eq.(5.185) will not exist. In fact, a solution to eq.(5.185) exists if and only if y belongs to the column space[8] of A. We can use an equation-error

[8] The column space of a matrix X is the linear vector space spanned by the columns of X.

minimization method to find a (hopefully) reasonable solution, even when eq.(5.185) itself has no solution, as we did in section 5.3.1. Our aim will be to understand why this solution may tend to be very sensitive to noise. This will motivate the discussion of section 5.7.

5.6.1 Column-Space Analysis

The value of x which we seek is that which minimizes:

$$J = ||Ax - y||^2 \tag{5.186}$$

where $||\cdot||$ is the euclidean norm for vectors. (Actually, other norms could also be used.)

Let us consider the nature of the solution of $\min_x J$. The extrema of J are found as:

$$\begin{aligned} 0 &= \frac{\partial J}{\partial x} && (5.187) \\ &= 2A^T(Ax - y) && (5.188) \end{aligned}$$

which implies:

$$A^T Ax = A^T y \tag{5.189}$$

Now, let cs(A) denote the column space of A. The vector y can be expressed as the sum of two vectors: $y_{\|}$ belonging to cs(A) and $y_{\perp}$ belonging to the complementary subspace:

$$y = y_{\|} + y_{\perp} \tag{5.190}$$

$y_{\perp}$ is orthogonal to the columns of A:

$$A^T y_{\perp} = 0 \tag{5.191}$$

So now, using eqs.(5.190) and (5.191), we can write eq.(5.189) as:

$$A^T Ax = A^T y_{\|} \tag{5.192}$$

or, equivalently:

$$A^T(Ax - y_{\|}) = 0 \tag{5.193}$$

By definition, $y_{\|}$ belongs to the column space of A, which means that $y_{\|}$ can be expressed as a linear combination of the columns of A. Hence a solution exists for:

$$Ax = y_{\|} \tag{5.194}$$

This vector x solves eq.(5.193) and will be the least-squares or minimum equation-error solution.

Now consider what happens when A and y are corrupted by noise, Δ and η respectively. Eq.(5.185) then becomes:

$$(A + \Delta)x = y + \eta \tag{5.195}$$

The least-squares solution of eq.(5.195) is obtained by solving the analog of eq.(5.194), which is:

$$(A+\Delta)x = y_{\|_{A+\Delta}} + \eta_{\|_{A+\Delta}} \tag{5.196}$$

where $y_{\|_{A+\Delta}}$ and $\eta_{\|_{A+\Delta}}$ are the projections (components) of y and η, respectively, in the column space of the corrupted matrix, $A+\Delta$.

It is clear that η, the error in y, will introduce error in the solution of eq.(5.196).

Δ, the error of A, may also have a subtle but substantial effect. We can see this as follows. Δ originates from errors. Thus, even though Δ may be of small magnitude, the structure of its columns is very different from the structure of the columns of A. For example, even though Δ and A may have zeros at the same elements, we would not usually expect columns of Δ to be proportional to columns of A. This means that the column space of $A+\Delta$ may be quite different from the column space of A alone, even though the norms of A and $A+\Delta$ may differ only marginally. But $y_{\|_{A+\Delta}}$ and $\eta_{\|_{A+\Delta}}$ are the projections of y and η into the column space of $A+\Delta$. Since $\text{cs}(A+\Delta)$ is possibly quite different from $\text{cs}(A)$, we expect that $y_{\|_{A+\Delta}}$ and $\eta_{\|_{A+\Delta}}$ may differ substantially from $y_{\|_A}$ and $\eta_{\|_A}$, the projections of y and η into the column space of the uncorrupted matrix, A. Thus the error matrix, Δ, may have small norm and yet still have a substantial effect in distorting the least-squares solution.

Continuing this qualitative discussion, let us note that the difference between $\text{cs}(A+\Delta)$ and $\text{cs}(A)$ involves the *dimension* of the space, $\Re^N$, of which these are subspaces. We should expect some proportionality between the disparity of these column spaces and the dimension, N. If N is large and $\text{rank}(A) \ll N$, then $\text{cs}(A+\Delta)$ has a lot of "room" into which to diverge from $\text{cs}(A)$. In this case, the distortion of the solution may be substantial. On the other hand, if $\Re^N$ is itself a space of low dimension then $\text{cs}(A)$ "fills" most of this space by itself, so $\text{cs}(A+\Delta)$ cannot differ from $\text{cs}(A)$ by too much. The consequence is that we should expect ill-conditioning to be more problematic in systems of high dimensionality than in low-dimensional systems.

Example 6 Let us consider a simple example of the effect of rank, dimension and uncertainty on the least-squares solution. Let e^k denote the kth standard basis vector in $\Re^N$: unity in the kth position, zero elsewhere. Let the matrix A be e^1:

$$A = \text{e}^1 \in \Re^{N\times 1} \tag{5.197}$$

and let y be a vector of ones: $y^T = (1, \ldots, 1)$. The projection of y into the column space of A is:

$$y_{\|} = y_1 \text{e}^1 = \text{e}^1 \tag{5.198}$$

Hence eq.(5.194) becomes:

$$\text{e}^1 x = \text{e}^1 \tag{5.199}$$

whose solution is:

$$x = 1 \tag{5.200}$$

Now add an uncertain term to the matrix A:

$$\Delta = \beta \sum_{k=2}^{N} \mathrm{e}^k, \quad \beta = \frac{1}{N-1} \tag{5.201}$$

The norm of Δ is $1/\sqrt{N-1}$, which is small if N is large. The projection of y into the column space of $A + \Delta$ becomes:

$$y_{\|A+\Delta} = \frac{2(N-1)}{N} \left(\mathrm{e}^1 + \Delta\right) \tag{5.202}$$

Now eq.(5.194) becomes:

$$\left(\mathrm{e}^1 + \Delta\right) x_{A+\Delta} = \frac{2(N-1)}{N} \left(\mathrm{e}^1 + \Delta\right) \tag{5.203}$$

whose solution is:

$$x_{A+\Delta} = \frac{2(N-1)}{N} \tag{5.204}$$

The fractional change in the least-squares solution, as a result of the small error in A is:

$$\frac{|x - x_{A+\Delta}|}{x} = \left|\frac{N-2}{N}\right| \tag{5.205}$$

which approaches unity as the dimension, N, increases, while the norm of the error matrix Δ becomes ever smaller. ■

5.6.2 Multiplicity of Solutions

Let us consider the multiplicity of solutions of eq.(5.185). Discussion of this question, directed specifically at linear elastic systems, is found in [66]. See also [56, section 3.10].

A necessary and sufficient condition for existence of solutions of eq.(5.185) is that A and the matrix $[A,\ y]$, formed by adjoining y as an additional column in A, have the same rank:

$$\mathrm{rank}[A] = \mathrm{rank}[A,\ y] \tag{5.206}$$

In other words, solutions exist for eq.(5.185) if and only if y is in the column space of A.

Let us now consider systems satisfying eq.(5.206). How many solutions does the system have, and what is the structure of these solutions?

Let r be the dimension of the column null space of A. In other words, there are exactly r linearly independent solutions of the homogeneous equation:

$$Ax = 0 \tag{5.207}$$

Let us denote these solutions $x^1, \ldots, x^r$. Let ns(A) denote the column null space of A.

Now, let x^0 be a particular solution of eq.(5.185):

$$Ax^0 = y \tag{5.208}$$

Every solution x of eq.(5.185) can be expressed as:

$$x = x^0 + x' \tag{5.209}$$

where $x' \in$ ns(A), because $Ax' = 0$. Furthermore, there are precisely r linearly independent solutions like eq.(5.209):

$$x^0 + x^1, \ldots, x^0 + x^r \tag{5.210}$$

In conclusion, if A is very rank deficient, then its null space has high dimension, r. Consequently, if *any* solutions of $Ax = y$ exist, then there are many linearly independent solutions. Note in particular that the multiplicity of solutions is explicitly a result of the dimensionality of the solution space. Hence the problem is again more severe in systems of high dimension than of low dimension.

5.7 Selective Sensitivity

High-dimensional ill-conditioning exists in many practical fault diagnosis problems. In this section we will study the reliability of a diagnostic system-identification method which partitions a large-dimensional problem into many low-dimensional ones, thus ameliorating the ill-conditioning to some extent. However, there is "no free lunch" in fault diagnosis, and this advantage is gained at the expense of extensive sensor requirements.

5.7.1 Basic Concept of Selective Sensitivity

The concepts and methods of selective sensitivity are discussed elsewhere [10, 13, 23, 29, 42, 76]. The aim here is only to make a brief operational review of the basic idea.

Consider an N-dimensional linear elastic system:

$$M\ddot{x}(t) + C\dot{x}(t) + Kx(t) = Hf(t) \tag{5.211}$$

where M, C and K are inertia, damping and stiffness matrices, respectively. The deflection vector is $x(t)$, the input vector is $f(t)$ and the measurement vector is $y(t)$:

$$y(t) = Gx(t) \tag{5.212}$$

G and H are real matrices, possibly rectangular, which indicate the locations of the sensors and actuators, respectively.

Let $\phi(s)$ and $\psi(s)$ represent the Laplace transforms of the input and the output vectors. Then eqs.(5.211) and (5.212) can be transformed and combined as:

$$\psi(s) = GF(s)H\phi(s) \tag{5.213}$$

where $F(s)$ is the transfer matrix (or frequency response matrix). Its inverse, $Q(s)$ (the dynamic stiffness matrix) is:

$$Q(s) = s^2 M + sC + K \tag{5.214}$$

The matrices M, C and K depend on physical model parameters such as stiffnesses, masses, geometrical parameters, etc., which we denote p_1, p_2, ... The output sensitivity to the nth model parameter, p_n, is defined here as the norm of the differential output variation to the nth parameter:

$$S(p_n) = \left\| \frac{\partial \psi(s)}{\partial p_n} \right\|^2 \tag{5.215}$$

where $\| \ \|$ is the euclidean norm for vectors.

When $S(p_n)$ is large, the Laplace-transformed output is sensitive to variations of the model parameter p_n, while if $S(p_n)$ is small then little information about p_n is contained in the measurements. A Laplace-transformed output which is sensitive to a given model parameter can be used to estimate that parameter, while an output which is insensitive to a model parameter will not be useful for estimating that parameter. A selectively sensitive input for p_n is an input which causes large sensitivity to p_n and small (hopefully vanishing) sensitivity to all other model parameters. The main goal of selective sensitivity is to seek a selectively sensitive input for each parameter (or for small groups of parameters) and thereby to decompose the overall system identification into a sequence of low-order estimation problems. Since ill-conditioning tends to be more severe for high- than for low-dimensional estimations, this decomposition may be expected to reduce the ill-conditioning and to enhance the robustness of the diagnosis.

We now state a sufficient condition for the existence of a selectively sensitive input. (This condition is both necessary and sufficient if G is square and invertible). This condition also indicates how such an excitation may be constructed. Various analytical, numerical and practical aspects of this condition are discussed in the references mentioned earlier.

One is often justified in assuming that the dynamic stiffness matrix, $Q(s)$, is a linear function of the model parameters, p_1, p_2, ... In other words, $Q(s)$ can be expressed as:

$$Q(s) = \sum_{n=1}^{N_p} p_n A_n(s) \tag{5.216}$$

The $N \times N$ matrices A_n depend on the topology of the structure — what is connected to what — and on the formulation of the finite element representation of the structure, but not on the parameter values themselves.

Once the finite element representation of the system has been chosen (not always a trivial matter in itself), the matrices A_n are completely known. For example, the mass, damping and stiffness matrices are often linear in local physical parameters ([68], [69]):

$$M = \sum_n m_n M_n\,, \quad C = \sum_n c_n C_n\,, \quad K = \sum_n k_n K_n \tag{5.217}$$

where the matrices M_n, C_n and K_n do not depend on the physical model parameters m_n, c_n and k_n which represent local mass, damping and stiffness.

Employing eq.(5.213) one finds the sensitivity to p_n to be the following hermitian quadratic form in the Laplace transform of the input:

$$S(p_n) = \phi^\dagger \underbrace{H^T[FA_nF]^\dagger G^T G[FA_nF]H}_{D_n} \phi \tag{5.218}$$

$$= \phi^\dagger D_n \phi \tag{5.219}$$

where (5.218) defines the sensitivity matrix, D_n. (Superscript T implies the matrix transpose; superscript † means matrix conjugate transposition.)

Let $\mathcal{I}$ be a set of indices of model parameters. As a matter of definition, the condition for selective sensitivity to the parameters indexed in $\mathcal{I}$ is:

$$S(p_n) = \begin{cases} 0 & \text{if} \quad n \notin \mathcal{I} \\ \text{not zero} & \text{if} \quad n \in \mathcal{I} \end{cases} \tag{5.220}$$

Combining this with relation (5.219), one can show that an input causes selective sensitivity to the model parameters whose indices are contained in the index-set $\mathcal{I}$ if vectors θ and ϕ exist which satisfy the following two relations. In fact, ϕ will be the desired selectively sensitive input.

$$A_n\theta = \begin{cases} 0 & \text{if} \quad n \notin \mathcal{I} \\ \text{not zero} & \text{if} \quad n \in \mathcal{I} \end{cases} \tag{5.221}$$

$$H\phi = Q\theta \quad \text{or} \quad FH\phi = \theta \tag{5.222}$$

It is important to note that solutions, θ, of eq.(5.221) are independent of the model parameters, and depend only on the topology of the system. In addition, solution vectors θ and ϕ for eqs.(5.221) and (5.222) exist for most parameter sets $\mathcal{I}$ if the topology matrices A_n are sparse. This is usually the case in elastic mechanical systems.

Exploiting the orthogonality properties of θ in eq.(5.221) and the linear sub-system modelling, eq.(5.216), the lefthand relation in eq.(5.222) becomes:

$$H\phi = \sum_{n\in\mathcal{I}} p_n A_n \theta \tag{5.223}$$

We now consider the physical interpretation of a selective input. An input exists which is selectively sensitive for parameter p_n if the output, θ, satisfies:

$$A_i\theta = \gamma\delta_{in} \tag{5.224}$$

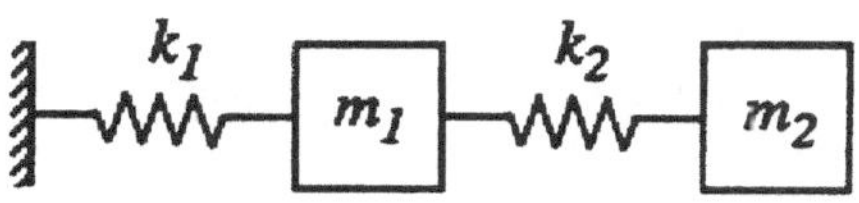

Figure 5.18: Undamped two-dimensional system.

where γ is a non-zero constant vector. If A_i is symmetric and positive (or negative) semi-definite, then eq.(5.224) is equivalent to:

$$\theta^T A_i \theta = \theta^T \gamma \delta_{in} \tag{5.225}$$

Now for the physical meaning. Suppose $A_i = K_i$, so $k_i K_i$ is a local stiffness matrix. Then $\theta^T A_i \theta$ is proportional to a local strain energy. Thus an input is selective for (against) the stiffness parameter k_n if and only if the resulting local strain energy is not zero (is zero).

5.7.2 Example: 2-Dimensional System

Consider the undamped 2-dimensional system shown in fig. 5.18: two masses connected in series by linear compressional springs; the system is fixed at one end, free at the other. The mass matrix is $M = \text{diag}(m_1, m_2)$. The stiffness matrix is:

$$K = \begin{pmatrix} k_1 + k_2 & -k_2 \\ -k_2 & k_2 \end{pmatrix} \tag{5.226}$$

Let us suppose that the masses m_1 and m_2 are known, and we wish to find a selectively sensitive excitation for determining k_1. That is, we seek an input such that the Laplace-transformed output is sensitive to k_1 and insensitive to k_2. We will find that this can be done with harmonic inputs, but actuators acting directly on both masses are necessary.

We seek an input for selectively sensitive estimation of k_1. Let us identify the four model parameters as:

$$p_1 = m_1, \quad p_2 = m_2, \quad p_3 = k_1, \quad p_4 = k_2 \tag{5.227}$$

We assume that p_1 and p_2 are known. The dynamic stiffness matrix, $Q(s)$, is linear in these model parameters, as in eq.(5.216), where the matrices A_n are:

$$A_1 = \begin{pmatrix} s^2 & 0 \\ 0 & 0 \end{pmatrix}, \quad A_2 = \begin{pmatrix} 0 & 0 \\ 0 & s^2 \end{pmatrix} \tag{5.228}$$

$$A_3 = \begin{pmatrix} 1 & 0 \\ 0 & 0 \end{pmatrix}, \quad A_4 = \begin{pmatrix} 1 & -1 \\ -1 & 1 \end{pmatrix} \tag{5.229}$$

The first condition for selective sensitivity for k_1, eq.(5.221), is existence of a vector θ satisfying:

$$A_3\theta \neq 0, \quad A_1\theta = A_2\theta = A_4\theta = 0 \tag{5.230}$$

Clearly, no dynamic measurement can distinguish k_1 from m_1 (a static measurement could, though, since then $s = 0$). However, we are assuming the masses to be known, so we can relinquish the requirement that $A_1\theta = A_2\theta = 0$. One can distinguish between the stiffness parameters by choosing θ as:

$$\theta(s) = \begin{pmatrix} 1 \\ 1 \end{pmatrix} \zeta(s) \tag{5.231}$$

where $\zeta(s)$ is an arbitrary scalar function. With this θ one finds $A_3\theta \neq 0$ and $A_4\theta = 0$.

Now, to actually find the input, ϕ, yielding selective sensitivity to k_1 and insensitivity to k_2, we must use this value of θ and solve eq.(5.222). In the present example, with two actuators, H is the identity matrix and eq.(5.222) becomes:

$$\phi = \begin{pmatrix} s^2 m_1 + k_1 \\ s^2 m_2 \end{pmatrix} \zeta(s) \tag{5.232}$$

We see then, that an input exists which causes the output to be sensitive to k_1 and insensitive to k_2, so the estimation of k_1 can be separated from the estimation of k_2. As always in selective sensitivity, the determination of this input depends on the model parameter being estimated (k_1) but not on the parameters to which the output is insensitive (k_2). We note that selective sensitivity for k_1 requires direct excitation of both masses: two actuators are needed.

5.7.3 Example: Structural Integrity of a Building

We will now consider an example with a larger model.

Substantial effort has been devoted to the design of seismic-resistent buildings and to the diagnosis of structural damage resulting from earthquake excitation [34, 84, 102, 107]. A major obstacle in assessing earthquake damage is the complexity of the structures involved and the resulting multiplicity of model parameters whose values must be up-dated.

There is a strong temptation to attempt this damage assessment on the basis of measurements of the building during the earthquake itself. This, however, is a difficult identification problem in all but the most severe cases, where direct inspection might reveal most of the damage anyway. The seismic input, applied only at ground level, is a poorly designed excitation for identifying localized damage in the structure. Fault detection in such circumstances is prone to be ill-conditioned.

The problem can be reduced by using the method of selective sensitivity. However, many actuators and sensors will be needed. We will illustrate this now.

The building will be modelled as a shear column. One can first estimate the stiffnesses by selectively sensitive static loads and deflections, and then use selectively sensitive dynamic excitations for determinating the masses

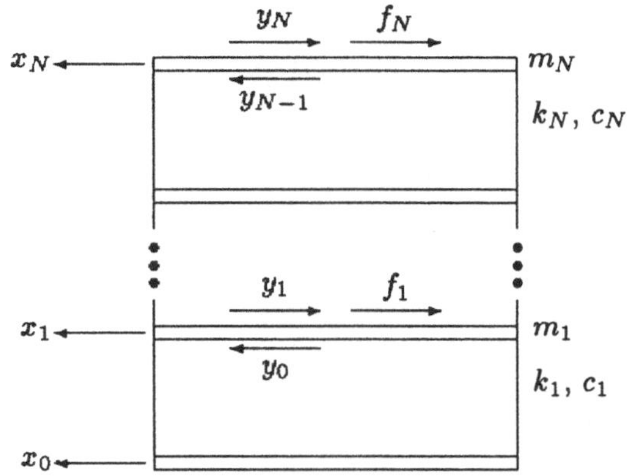

Figure 5.19: Variables and parameters of the shear-type building.

and damping coefficients. The deflection-equations are based on the work of Zhang and Soong [107].

The Model

The dynamic variables and model parameters of the shear-type building model are shown in fig. 5.19. N floor units are depicted. The lateral shear displacement of the nth floor unit is $x_n(t)$, while the resulting shear force in the columns of the nth floor unit is $y_{n-1}(t)$. External lateral loading on the nth floor is represented by $f_n(t)$. The mass of the nth floor is m_n and the stiffness and damping coefficients of the columns are k_n and c_n, respectively. The basic dynamic equations, based on [107], are:

$$f_n + y_n = y_{n-1} + m_n \ddot{x} \quad (5.233)$$
$$y_{n-1} = k_n[x_n - x_{n-1}] + c_n[\dot{x}_n - \dot{x}_{n-1}] \quad (5.234)$$

There is no ground motion during the integrity tests, so $x_0 = \dot{x}_0 = 0$. In addition, the shear at the top of the building must vanish: $y_N = 0$. The shear forces can be eliminated and these equations can be combined in matrix form as in eq.(5.211) with H as the identity matrix and with $M = \text{diag}(m_1, \ldots, m_N)$. The stiffness matrix $K = [k_{m,n}]$ is tridiagonal, with the following structure:

$$\begin{aligned} k_{n,n} &= k_n + k_{n+1} \\ k_{n,n+1} &= -k_{n+1} \\ k_{n,n-1} &= -k_n \end{aligned} \quad (5.235)$$

The damping matrix C has precisely the same form, with each k replaced by c. We adopt the convention that $k_{N+1} = c_{N+1} = 0$.

Let $\psi(s)$ and $\phi(s)$ be the Laplace transforms of $x(t)$ and $f(t)$, respectively. Then the Laplace transform of the equations of motion becomes:

$$Q(s)\psi(s) = \phi(s) \quad \text{or} \quad \psi(s) = F(s)\phi(s) \tag{5.236}$$

which is precisely eq.(5.213) with $G = H = I$.

The mass, damping and stiffness matrices can be represented in linear sub-model parametrizations as in eq.(5.217). Concise representation of these matrices is helpful in finding the excitations which are needed for inducing selectively sensitive response. Let e^γ represent the N-dimensional standard basis vector: unity in the γth position and zero elsewhere. Then one finds the following relations:

$$M_n = s^2 \mathrm{e}^n \mathrm{e}^{nT} \tag{5.237}$$

$$K_n = \left(\mathrm{e}^n - \mathrm{e}^{n-1}\right)\left(\mathrm{e}^n - \mathrm{e}^{n-1}\right)^T \tag{5.238}$$

$$C_n = sK_n \tag{5.239}$$

where the dynamic stiffness matrix is:

$$Q(s) = \sum_n \left[m_n M_n + c_n C_n + k_n K_n\right] \tag{5.240}$$

Conditions for Selective Sensitivity

We are now prepared to determine the conditions for selective sensitivity for the stiffness, mass and damping model parameters. We will determine which parameters can be selectively probed and what the inputs must be. We will find that a high degree of selectivity is possible, implying good conditioning of the estimation and low sensitivity to noise. However, many sensors and actuators will be required.

We begin by considering static loading to determine the shear stiffnesses. When using using static measurements the sensitivities to mass and damping are automatically zero. This comes about algebraically simply because $s = 0$. Condition (5.221) for selective sensitivity for the γth stiffness is existence of an output vector θ such that:

$$K_n \theta = \begin{cases} 0, & n \neq \gamma \\ \text{not zero}, & n = \gamma \end{cases} \tag{5.241}$$

One finds that

$$\theta = h\mathrm{e}^i \tag{5.242}$$

results in sensitivity to parameters indexed $n = i$ and $n = i - 1$, where h is a constant and e^i is a standard unit vector. Starting at the first storey ($n = 1$)

and proceeding upwards, one can estimate the stiffness coefficients one at a time.

The load required to achieve the output vector (5.242) is:

$$\phi = K\theta \tag{5.243}$$

$$= h\sum_{n=1}^{N} k_n K_n e^i \tag{5.244}$$

$$= -h\left(e^{i+1} - 2e^i + e^{i-1}\right) \tag{5.245}$$

A triplet of load points is needed for each stiffness diagnosis, if the stiffness coefficients are to be tested separately. While many actuators and sensors are needed, the stiffness estimation of this N-storey building is partitioned into N decoupled one-dimensional estimations. In light of our discussion in section 5.6, this will tend to reduce the ill-conditioning and enhance the reliability of the diagnosis.

We now inquire if we can obtain selective sensitivity to the γth mass. Using the fact that $FQ = I$, one finds that:

$$\partial F/\partial m_n = -F\partial Q/\partial m_n F \tag{5.246}$$

Hence the variation of the output ψ with the nth mass is:

$$\frac{\partial \psi}{\partial m_n} = -F\frac{\partial Q}{\partial m_n}F\phi \tag{5.247}$$

$$= -FM_nF\phi \tag{5.248}$$

Similarly, the output variations with respect to the stiffness and damping coefficients are:

$$\frac{\partial \psi}{\partial k_n} = -FK_nF\phi\,, \quad \frac{\partial \psi}{\partial c_n} = -FC_nF\phi \tag{5.249}$$

For selective sensitivity to m_γ we require:

$$S(m_n) = \begin{cases} 0 & n \neq \gamma \\ \text{not zero} & n = \gamma \end{cases} \tag{5.250}$$

and

$$S(k_n) = S(c_n) = 0\,, \quad n = 1, \ldots, N \tag{5.251}$$

Employing eq.(5.247) and (5.249), this is equivalent to:

$$M_nF\phi = \begin{cases} 0 & n \neq \gamma \\ \text{not zero} & n = \gamma \end{cases} \tag{5.252}$$

and

$$K_nF\phi = C_nF\phi = 0\,, \quad n = 1, \ldots, N \tag{5.253}$$

But $\psi = F\phi$, so a necessary condition for selective sensitivity to the γth mass is that the output satisfy the following relations:

$$M_\gamma \psi \neq 0 \tag{5.254}$$
$$M_n \psi = 0\,, \quad n \neq \gamma \tag{5.255}$$
$$C_n \psi = 0\,, \quad n = 1, \ldots, N \tag{5.256}$$
$$K_n \psi = 0\,, \quad n = 1, \ldots, N \tag{5.257}$$

Using eq.(5.237), one finds that eqs.(5.254) and (5.255) imply that $\psi(s)$ must be proportional to the γth standard basis vector:

$$\psi(s) = h(s)e^\gamma \tag{5.258}$$

where $h(s)$ is the Laplace transform of an arbitrary (non-zero) scalar function. This choice of ψ then violates eqs.(5.256) and (5.257) for $n = \gamma$ and $n = \gamma+1$. In other words, sensitivity to a single mass parameter, m_γ, is invariably accompanied by sensitivity to the damping and stiffness parameters c_γ, $c_{\gamma+1}$, k_γ and $k_{\gamma+1}$. The response will, however, be insensitive to all of the remaining model parameters.

The Laplace transform of the load required to produce the desired selectivity is found as follows from eqs.(5.236) and (5.258):

$$\phi(s) = Q(s)\psi(s) \tag{5.259}$$
$$= h(s)Q(s)e^\gamma \tag{5.260}$$
$$= h(s)\left[m_\gamma M_\gamma + c_\gamma C_\gamma + c_{\gamma+1}C_{\gamma+1} + k_\gamma K_\gamma + k_{\gamma+1}K_{\gamma+1}\right]e^\gamma \tag{5.261}$$
$$= h(s)\left[(s^2 m_\gamma + sc_\gamma + sc_{\gamma+1} + k_\gamma + k_{\gamma+1})e^\gamma - (sc_\gamma + k_\gamma)e^{\gamma-1} - (sc_{\gamma+1} + k_{\gamma+1})e^{\gamma+1}\right] \tag{5.262}$$

One sees that selective sensitivity for m_γ and c_γ is obtained by dynamically loading the building at three points only: at the $(\gamma-1)$th, γth and $(\gamma+1)$th floors. (When estimating m_1 and c_1 only two external loads are required). This triplet of actuators must move along the structure as successive floors are tested. Again we expect enhanced reliability and reduced ill-conditioning of the parameter estimation, at the cost of many sensors and actuators.

5.8 Problems

1. *Response set for the bending moment.* Consider the simply supported uniform beam in fig. 5.1, which carries an uncertain load $\phi(x)$ distributed over an unknown length $y \leq L/2$. The uncertainty of the load profile is represented by a Fourier ellipsoid-bound model:

$$\phi(x) = \sum_{n=1}^{N} \beta_n \cos n\pi x/y = \beta^T \gamma(x) \tag{5.263}$$

where β is the Fourier-coefficient vector and $\gamma(x)$ is a vector of trigonometric functions. The uncertainty in β is represented:

$$\mathcal{U}(y) = \left\{\beta : \ (\beta - \overline{\beta})^T W (\beta - \overline{\beta}) \leq \alpha^2\right\} \tag{5.264}$$

Construct the response set for the bending moment measured at the midpoint of the beam.

2. *Identification of the nominal load.* Consider the simply supported uniform beam in fig. 5.1, which carries an uncertain load $\phi(x)$ distributed over a known length $y \leq L/2$. The uncertainty in the load profile is represented by eq.(5.2). Use the moment measured at the midpoint to diagnosis the nominal load, $\overline{\phi}$. (a) Determine the fractional resolution of $\overline{\phi}$. (b) The diagnosis is satisfactory if the fractional resolution is no greater than the critical value, f_{cr}. Determine the robust reliability of the diagnosis.

3. *Multi-hypothesis diagnosis of a torque load.* Consider a rotating system subject to uncertain transient torque loads. The dynamics are modelled as:

$$J\ddot{\theta}(t) + k\theta(t) = u(t), \quad \theta(0) = \dot{\theta}(0) = 0 \tag{5.265}$$

The uncertain torque belongs to one of two convex models, one corresponding to small transients and the other to large transients:

$$\mathcal{U}_i(\alpha) = \left\{u(t) : \int_0^\infty (u(t) - \overline{u}_i(t))^2 \, dt \leq \alpha^2\right\}, \quad i = 1, 2 \tag{5.266}$$

where $0 < \overline{u}_1(t) < \overline{u}_2(t)$, so $\mathcal{U}_1$ and $\mathcal{U}_2$ are sets of uncertain small and large transients, respectively. Let $\overline{\theta}_i(t)$ represent the angular displacement in response to $\overline{u}_i(t)$. The angular displacement θ is measured at one instant, and the following nearest-neighbor rule is used to decide between the two convex models. Decide that the load came from $\mathcal{U}_i$ if:

$$(\theta - \overline{\theta}_i)^2 < (\theta - \overline{\theta}_j)^2 \tag{5.267}$$

This decision algorithm is considered to succeed if it invariably identifies the correct input set. What is the robust reliability of the algorithm? That is, how much uncertainty can the algorithm tolerate without failure?

4. ‡ *Modification of problem 3.* Measure the angular displacement at N instants: $t_1 < \cdots < t_N$. Let $\overline{\theta}_i(t)$ represent the vector of angular displacements in response to $\overline{u}_i$ at these times, and let θ_u denote the vector of displacements for arbitrary load $u(t)$. We use a nearest-neighbor decision rule like relation (5.267), except now we consider the euclidean norm of the difference. Choose $\mathcal{U}_i$ if:

$$\left\|\theta - \overline{\theta}_i\right\|^2 < \left\|\theta - \overline{\theta}_j\right\|^2 \tag{5.268}$$

What is the robust reliability of this diagnosis with N measurements?

5. *Multi-hypothesis input diagnosis.* Consider a two-dimensional system:

$$\dot{x} = ax(t) + b(t) + \beta(t), \quad \dot{x}(0) = 0 \tag{5.269}$$

where a is a scalar, $x(t)$ is a 2-dimensional state vector, $b(t)$ is a known input, and $\beta(t)$ is an unknown anomalous input. The response is denoted $x_\beta(t)$.

Let the anomalous input belong to one of two sets of functions:

$$\mathcal{F}_1(\alpha) = \left\{\beta(t) : \ [\beta_1(t) - \overline{\beta}]^2 \le \alpha^2, \quad \beta_2^2(t) \le \alpha^2\right\} \tag{5.270}$$

$$\mathcal{F}_2(\alpha) = \left\{\beta(t) : \ [\beta_2(t) - \overline{\beta}]^2 \le \alpha^2, \quad \beta_1^2(t) \le \alpha^2\right\} \tag{5.271}$$

We consider two hypothesized modifications, one from each of these sets:

$$h_1 = \begin{pmatrix} \overline{\beta} \\ 0 \end{pmatrix} \in \mathcal{F}_1, \quad h_2 = \begin{pmatrix} 0 \\ \overline{\beta} \end{pmatrix} \in \mathcal{F}_2 \tag{5.272}$$

Can we distinguish the two classes of modifications, $\mathcal{F}_1$ and $\mathcal{F}_2$, by comparing the measured response, $x(t)$, at a particular instant, against the anticipated response from either h_1 or h_2? The decision algorithm is that we choose $\mathcal{F}_n$ if:

$$\|x - x_{h_n}\| = \min_{m=1,2} \|x - x_{h_m}\| \tag{5.273}$$

What is the robust reliability of this algorithm? That is, what is the greatest value of the uncertainty parameter, α, such that the decision is always correct, for any modification in either $\mathcal{F}_1$ or $\mathcal{F}_2$?

6. *Diagnosis of stiffness.* The body and trailer of a two-stage autobus are connected by heavy springs which allow limited vertical in-plane rotation between the two sections. A vibration sensor monitors the angular deflection $\theta(t)$, and an on-board microprocessor evaluates the mean-squared deflection:

$$\overline{\theta^2} = \frac{1}{T}\int_0^T \theta^2(t)\, dt \tag{5.274}$$

The measurement is used to detect anomalous decrease in the rotational stiffness. Assume the rotational vibration is described by a single degree of freedom, as in eq.(5.265). Prior information about the uncertain input $u(t)$ establishes a Fourier-ellipsoid bound convex model:

$$\mathcal{U}(\alpha) = \left\{u(t) = \beta^T\gamma(t) : \ \beta^T W \beta \le \alpha^2\right\} \tag{5.275}$$

where $\gamma(t)$ is a known vector of trigonometric functions. The nominal stiffness is k_0, and the measured value of $\overline{\theta^2}$ is used to distinguish k_0 from k_1. (a) What is the robust reliability of the diagnosis? (b) For a given amount of uncertainty, α, what is the smallest fractional increment of stiffness which can be unambiguously diagnosed?

Chapter 6

Reliability of Mathematical Models

6.1 Models, Decisions and Reliability

Models of mechanical systems are developed for various purposes, including design, safety assessment, dynamic analysis and so on. No model is perfect, and model inaccuracy reduces the reliability of decisions based on the model. Furthermore, there is no unique definition of the accuracy of a model. Rather, model inaccuracy should be evaluated with respect to the intended use of the model. In this chapter we will evaluate the reliability of models in terms of the robustness-to-uncertainty of decisions based on the model.

The concept of robust model reliability developed here is that a model is reliable if subsequent applications of the model can tolerate a large amount of uncertainty. Conversely, model-based decisions which are fragile with respect to uncertainty, which can tolerate only a small amount of uncertainty before unacceptable or unstable decisions become possible, are unreliable.

Before we discuss some examples, let us outline our analysis of model reliability. Our models will have two components. The *mechanical model* is based on engineering information and physical principles, and describes some properties of the system. For example the differential equation describing flexural vibration of a beam is a mechanical model. Or, the spatial variation of a material property such as stiffness may sometimes be the relevant mechanical model. The *uncertainty model* quantifies the possible inaccuracy of the mechanical model as well as the uncertainty of the environment within which the system operates, such as load uncertainty. We will use convex models to represent uncertainty. The final link in our analysis is the *decision* or application which is based on the mechanical model. The mechanical model is reliable and the decision is robust if the decision does not change significantly as the model varies over the set of possible values consistent with the

System Reliability		**Model Reliability**
Mechanical System	$\Longrightarrow$	Model and Decision
Uncertainty Models	$\Longrightarrow$	Uncertainty Models
Failure Criterion	$\Longrightarrow$	Decision Stability

Table 6.1: Analogy between robust reliability of systems and of mathematical models.

uncertainty.

This concept of the robust reliability of mathematical models is a direct analog of robust system reliability. This is certainly not the only way one could define the reliability of a mathematical model. It can be useful to talk of a model as being reliable if we know that it very accurately predicts particular behavior. In this case we would be defining reliability in terms of small uncertainties associated with that model. Decisions based on such a model might not be robust to uncertainties, but this would not be important, since the uncertainties are known to be small. Such a model does not have the property of robust reliability which we will develop in this chapter.

The analogy between robust reliability of mechanical systems, which we have studied in chapters 3 and 4, and the robust reliability of mathematical models, is summarized in table 6.1. We replace the physical system by the mathematical model together with the subsequent decisions based on that model, and we consider the stability or acceptability of those decisions rather than the mechanical failure of the system.

We begin with two examples, to illustrate some of the possible implementations of this sort of reliability analysis. In section 6.2 we will consider the reliability for safety-assessment of an uncertain geometrical model of a cooling fin subject to fluid loading. In section 6.3 we study the reliability in predicting the dynamic behavior of a large-dimensional elastic structure which is approximately represented with a limited number of modal degrees of freedom. In section 6.4 we consider the more general problem of robust identification based on multi-hypothesis testing.

6.2 Cooling Fin With Uncertain Geometry

Let us consider a cooling fin of length L and tilted at an angle θ with respect to the flow direction of a fluid which exerts a force of ϕ [Newton/vertical meter] on the fin.[1] The fin has rectangular cross section with uniform width W and variable but uncertain thickness $2T(v)$, $0 \leq v \leq L$. The material

[1] We analyzed the reliability of this system with respect to uncertain fluid loading in section 3.3, pp.39–47.

properties are such that the fin will fail if the normal tensile stress at any point on the fin cross section exceeds the critical value s_{cr}.

The *model* we wish to develop is the thickness profile, $T(v)$, and the question we ask is: how accurately do we need to measure this profile? The *uncertainty* in the model — the uncertainty in $T(v)$ — will be expressed by a convex model $\mathcal{U}(\alpha)$. The *decision* to be made on the basis of the model is: will the fin break under a given load ϕ? The question of robustness is: for a given model, $T(v)$, how much uncertainty or inaccuracy can be tolerated in the model without altering the decision? The model is reliable if the decision is unaltered even if the model is quite erroneous.

The normal stress in the cross section at position v along the fin is greatest at the outer surface, a distance $T(v)$ from the normal plane, and equals:

$$s(v) = \frac{M(v)T(v)}{I(v)} \tag{6.1}$$

where $M(v) = 0.5\phi(L-v)^2 \sin^2\theta$ is the bending moment in the fin at position v and $I(v) = 2WT^3(v)/3$ is the area moment of inertia of the cross section. Using these expressions for M and I, the maximum tensile stress in the cross section at position v is:

$$s(v) = \frac{3\phi(L-v)^2 \sin^2\theta}{4WT^2(v)} \tag{6.2}$$

Let $\tau(v)$ denote the unknown model error at position v along the fin, and represent the model uncertainty by the convex model:

$$\mathcal{U}(\alpha) = \left\{ \tau(v) : \left| \frac{\tau(v)}{T_0(v)} \right| \leq \alpha \right\} \tag{6.3}$$

where $T_0(v)$ is the model which is adopted. In otherwords, the fractional error of the model at any position along the fin can be as large as α. The parameter α is a measure of the degree of uncertainty of the model. The model is reliable if the decision regarding the safety of the fin will not change even if α is large. So our analysis concentrates on the question: how large can α be without altering the safety-decision?

The fin will not break if the tensile strength is not exceeded anywhere along its length, so the decision is based on comparing the maximum tensile strength based on the model, $T_0(v)$, against the critical value of the strength:

$$s_{cr} \geq \frac{3\phi(L-v)^2 \sin^2\theta}{4WT_0^2(v)} \tag{6.4}$$

The fin is "safe" if this inequality is satisfied for all $0 \leq v \leq L$, but "unsafe" if the opposite inequality holds anywhere along the fin.

The uncertainty in the model at position v is $\tau(v)$. The greatest uncertainty which does not alter the decision is evaluated by treating (6.4) as an equality and replacing T_0 by $T_0 + \tau$:

$$s_{cr} = \frac{3\phi(L-v)^2 \sin^2\theta}{4W[T_0(v) + \tau(v)]^2} \tag{6.5}$$

The thickness $T_0 + \tau$ is non-negative, so the uncertainty satisfies $-T_0 \leq \tau$, as well as belonging to the convex model $\mathcal{U}(\alpha)$. Thus the critical uncertainty, which just causes the decision to switch, is obtained by rearranging (6.5) to yield:

$$\frac{\tau(v)}{T_0(v)} = -1 + \left(1 - \frac{v}{L}\right) \frac{L \sin\theta}{2T_0(v)} \sqrt{\frac{3\phi}{s_{cr} W}}, \quad 0 \leq v \leq L \tag{6.6}$$

But the uncertainty model, $\mathcal{U}(\alpha)$, requires that:

$$\left| \frac{\tau(v)}{T_0(v)} \right| \leq \alpha \tag{6.7}$$

Thus the greatest model uncertainty which leaves the safety-decision unchanged is:

$$\widehat{\alpha} = \min_{0 \leq v \leq L} \left| -1 + \left(1 - \frac{v}{L}\right) \frac{L \sin\theta}{2T_0(v)} \sqrt{\frac{3\phi}{s_{cr} W}} \right| \tag{6.8}$$

$\widehat{\alpha}$ is the greatest value of the uncertainty parameter which leaves the decision insensitive to the model uncertainty. This is the "robustness" of the decision or the "robust reliability" of the model. When $\widehat{\alpha}$ is large the decision is robust with respect to uncertainty and the model is reliable; when $\widehat{\alpha}$ is small the decision is fragile and the model is not reliable. $\widehat{\alpha}$ is a dimensionless ratio of thicknesses, and represents the relative error with which we must measure the thickness profile in order to assure a stable decision.

Let us examine the special case of constant nominal thickness, T_0. Define:

$$\beta = \frac{L \sin\theta}{2T_0} \sqrt{\frac{3\phi}{s_{cr} W}} \tag{6.9}$$

From eq.(6.6), the absolute value of the fractional error of the model can be expressed:

$$\left| \frac{\tau(v)}{T_0} \right| = \left| -1 + \left(1 - \frac{v}{L}\right) \beta \right| \tag{6.10}$$

which is shown in fig. 6.1. From these figures and eq.(6.8) we see that the robustness of the decision, $\widehat{\alpha}$, is zero if $\beta \geq 1$, and decreases linearly with β if $\beta \leq 1$:

$$\widehat{\alpha} = \begin{cases} 1 - \beta & \beta \leq 1 \\ 0 & \beta \geq 1 \end{cases} \tag{6.11}$$

In other words, when β exceeds unity, the reliablity of the model vanishes: arbitrarily small error in the nominal model can cause the decision about the

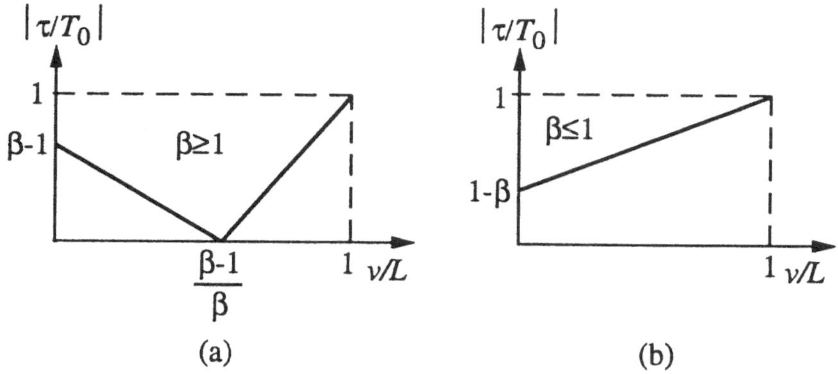

Figure 6.1: $\left|\frac{\tau(v)}{T_0}\right|$ versus β, for $\beta \geq 1$ (a) and $\beta \leq 1$ (b).

safety of the fin to switch from "yes" to "no" or the reverse. In other words, when $\beta \geq 1$ the decision is not robust in the sense of being quite fragile with respect to modelling error. On the other hand, when $\beta < 1$ the decision can tolerate some model uncertainty and still remain stable. The smaller the value of β, the more reliable is the model.

β is a function of physical parameters, so the dependence of the robustness on β can lead to physical insight and also has design implications. For example, since β increases with the square root of the force density ϕ, the robust reliability $\widehat{\alpha}$ of the model decreases with $\sqrt{\phi}$ up to a critical value (at which $\beta = 1$) and thereafter $\widehat{\alpha} = 0$, indicating no robustness of the decision to model uncertainty. Conversely, β decreases inversely with the square root of the fin width, indicating that wider fins will be more reliable in terms of the stability of the safety decision.

6.3 Modal Truncation of a High-Dimensional Model

In many industrial design and analysis applications one develops a numerical finite-element model for describing small-amplitude vibrations of a structure. A complex object, like a car body or a bridge, has a vast number of modes of vibration. The accurate description of the possible vibrations of such a system requires a model of very high dimension, so these finite-element models can have many thousands of degrees of freedom.

In the design stage, a preliminary model is constructed based on physical principles and prior experience. However, when the structure is built it is usually necessary to improve the preliminary model by performing measurements, so that the model can be used for accurate prediction and final design and safety analyses.

A common approach in measuring large elastic structures is to measure the frequency and shape of natural modes of vibration. However, it is rarely practical to measure all the modes of a large structure, nor is it necessary. Reliable decisions can be based on models obtained from measuring a limited number of natural modes. The questions are: how many and which modes should be measured? In this section we will study this question from the point of view of robust reliability.

Consider a large N-dimensional linear elastic structure subject to uncertain loads. We will model this system with only r of its N natural modes. We will then evaluate the reliability of the truncated model, in comparision with the full-rank model, in predicting displacements resulting from the uncertain inputs. We require that the prediction error not exceed a specified threshold. The maximum prediction error increases with both the input uncertainty and with the extent of modal truncation. This allows us to determine the greatest possible economy in the modal measurements, consistent with a given amount of input uncertainty and a specified acceptable prediction error. Or conversely, the robust reliability of a truncated model is the greatest value of the uncertainty consistent with specified acceptable prediction error. We will also be able to determine which modes should be retained in the reduced model.

In this example the *mechanical model* whose reliability is evaluated is the particular selection of the r modes we retain, from among the full set of N modes of the complete finite-element model. The *uncertainty model* is the input uncertainty model, which will be specified as a convex model. The *decision* in this case is to accept or reject the modal truncation of order r, depending on the magnitude of the maximum prediction error compared against the acceptable error threshold.

In the absence of damping the dynamics of the full model are described by:

$$M\ddot{x} + Kx = Bu \tag{6.12}$$

where $x(t)$ and $u(t)$ are the displacement and load vectors. The N normal mode vectors $\phi_1, \ldots, \phi_N$, and the corresponding eigenvalues $\omega_1^2, \ldots, \omega_N^2$, satisfy the eigenvalue equation:

$$\left(-\omega_n^2 M + K\right) \phi_n = 0 \tag{6.13}$$

The eigenvectors are normalized as $\phi_i^T M \phi_j = m_i \delta_{ij}$. The modal matrix is $\Phi = [\phi_1, \ldots, \phi_N]$. Define $V(t) \in \Re^{N\times N}$ as the diagonal matrix whose ith element is $(1/m_i\omega_i)\sin\omega_i t$. The displacement vector in response to load $u(t)$, for zero initial conditions, can be expressed:

$$x(t) = \int_0^t \Phi V(t-\tau)\Phi^T Bu(\tau)\,\mathrm{d}\tau \tag{6.14}$$

Let $S \in \Re^{N\times N}$ be a diagonal "selector" matrix with ones at positions corresponding to the indices of the r modes included in the modally truncated model and zeros elsewhere. The response of the modally-truncated model can be represented as:

$$\xi(t) = \int_0^t \Phi S V(t-\tau) S \Phi^T B u(\tau)\,\mathrm{d}\tau \tag{6.15}$$

The displacement at a particular node or along a particular direction is evaluated as the projection of the displacement vector along a constant unit vector ψ. Combining eqs.(6.14) and (6.15), the prediction error of the truncated model in comparison to the full model is:

$$\psi^T[x(t)-\xi(t)] = \psi^T \int_0^t \Phi \underbrace{[V(t-\tau) - SV(t-\tau)S]}_{\Delta(t-\tau)} \Phi^T B u(\tau)\,\mathrm{d}\tau \tag{6.16}$$

which defines the matrix Δ.

We consider a cumulative energy-bound convex model to describe the input uncertainty:

$$\mathcal{U}(\alpha) = \left\{ u(t) : \int_0^\infty u^T(t) u(t)\,\mathrm{d}t \le \alpha^2 \right\} \tag{6.17}$$

$\mathcal{U}(\alpha)$ is the set of load vectors whose integral-square values do not exceed α^2. The parameter α determines the size of the convex model and the uncertainty of the input functions.

The prediction error of the truncated model must not exceed the threshold value of $\Delta_{\rm cr}$:

$$\left|\psi^T[x(t)-\xi(t)]\right| \le \Delta_{\rm cr} \tag{6.18}$$

The Cauchy and Cauchy-Schwarz inequalities are employed to determine the extremal values of the prediction error, resulting in:

$$\max_{u\in\mathcal{U}(\alpha)} \psi^T[x(t)-\xi(t)] = \alpha \sqrt{\int_0^t \psi^T \Phi \Delta(\tau) \Phi^T B B^T \Phi \Delta(\tau) \Phi^T \psi^T\,\mathrm{d}\tau} \tag{6.19}$$

Combining the performance criterion, (6.18), with the maximum prediction error, (6.19), and solving for α, results in the greatest allowable uncertainty:

$$\widehat{\alpha} = \frac{\Delta_{\rm cr}}{\sqrt{\int_0^t \psi^T \Phi \Delta(\tau) \Phi^T B B^T \Phi \Delta(\tau) \Phi^T \psi^T\,\mathrm{d}\tau}} \tag{6.20}$$

The rank-r model has acceptable prediction error so long as the input uncertainty is no greater than $\widehat{\alpha}$. Decisions based on the rank-r model are "robust", immune to input uncertainty, up to uncertainty level $\widehat{\alpha}$. Thus $\widehat{\alpha}$ measures the robustness of the decision and the reliability of the reduced-rank model.

The denominator of eq.(6.20) increases monotonically in time, indicating that the reliability of the model decreases with the duration of operation of the system. For short duration the prediction-error is robust to large input uncertainty, while for long duration the model is less reliable.

Let us examine eq.(6.20) for the special case that $B = M = I$. Define $S_c = I - S$, and note that $S_c^2 = S_c$. Thus

$$\Delta^2 = [V - SVS]^2 = [S_c V S_c]^2 = S_c V^2 S_c \tag{6.21}$$

The robustness, eq.(6.20), becomes:

$$\widehat{\alpha} = \frac{\Delta_{\rm cr}}{\sqrt{\int_0^t \psi^T \Phi S_c V^2 S_c \Phi^T \psi^T \,{\rm d}\tau}} \tag{6.22}$$

If the model has full rank, $r = N$, then S_c vanishes and $\widehat{\alpha} = \infty$, indicating complete robustness of the prediction error to input uncertainty. As the rank decreases, S_c becomes fuller and the denominator of eq.(6.22) increases, reducing the robust reliability of the model.

6.4 Robust Multi-Hypothesis System Identification

Engineering decisions are often based on mathematical models, and in the previous sections we have evaluated the reliability of such models in terms of the robustness of the consequent decisions. But models themselves are frequently obtained through a process of measurement and decision. There are a plethora of methods and philosophies for system identification, and their study is beyond our scope. We will examine the robust reliability of one particular approach: system identification based on multi-hypothesis testing. We have considered the reliability of specific multi-hypothesis algorithms in sections 5.2 and 5.4. We now address the generic problem of evaluating the reliability of a model constructed by multi-hypothesis testing.

In robust reliability we ask the following question: how much uncertainty can be tolerated while still assuring acceptable performance? The *robust reliability* or *robustness* is a measure of the maximum acceptable uncertainty. Applying this to system identification, the robust reliability of an identification algorithm is a measure of the maximum uncertainty which is consistent with acceptable performance of the algorithm. The algorithm is reliable if it can tolerate a large amount of uncertainty; it is not reliable if it is fragile with respect to uncertainty.

In this section we will study the robust reliability of multi-hypothesis identification of a dynamic system for which empirical input/output data are available. The aim is to enable the evaluation and optimization of the model-updating procedure.[2]

[2]The author acknowledges with pleasure many stimulating conversations on this subject with Prof. Dr. H. G. Natke, director, and with Dr. Uwe Prells, Curt-Risch Institute for Dynamics, Sound, and Measurement, University of Hannover, Germany.

6.4.1 System Formulation

The system we wish to model is characterized by an N-dimensional state vector, $x(t)$, and can be excited by an N-dimensional input vector, $f(t)$. We have a partially satisfactory initial model for this system. Quite often the preliminary model is a linear approximation to the actual dynamic behavior, so we will denote the preliminary model by $\widehat{L}$.

The dynamic behavior of the system is precisely described by the following vector equation:

$$L(x, \dot{x}, \ddot{x}, t) + f(t) = h(x, \dot{x}, \ddot{x}, t) \tag{6.23}$$

L is that part of the system which we attempt to describe with the preliminary model $\widehat{L}$, while h represents those features of the system which we have not yet identified. For instance, a common circumstance in vibration dynamics is that L corresponds to linear vibration dynamics while h arises from non-linearities which frequently arise in joints and boundaries.

Our aim is to identify the operator h. We are able to perform measurements on the system,[3] so that for any given input $f(t)$ we can obtain the state vector, $x(t)$, and its derivatives, $\dot{x}(t)$ and $\ddot{x}(t)$. We can then calculate the approximation, $\widehat{L}(x, \dot{x}, \ddot{x}, t)$, to the lefthand part of the model, L, for that input.

6.4.2 Uncertainty in the Nominal Model

The nominal model $\widehat{L}$ is inaccurate, as expressed by an equation-error term ε when L in eq.(6.23) is replaced by $\widehat{L}$:

$$\widehat{L}(x, \dot{x}, \ddot{x}, t) + f(t) = h(x, \dot{x}, \ddot{x}, t) + \varepsilon(t) \tag{6.24}$$

The uncertainty in the nominal model is represented by a convex model for the error vector ε. Many different convex models are available, depending on the type of information available.

One convenient and common convex model is the instantaneous energy-bound model. At each instant, the weighted euclidean norm of the error is bounded:

$$\mathcal{U}(\alpha) = \{\varepsilon : \ \varepsilon^T A \varepsilon \leq \alpha^2\} \tag{6.25}$$

where A is a real, symmetric positive definite matrix. α is the uncertainty parameter, controlling the size of the convex model and quantifying the amount of uncertainty in the calculational model.

Another convex model, which we will consider in problem 5, is the integral energy-bound model. While the model of eq.(6.25) constrains the "energy" of the error at each instant, we now constrain the integral of the energy:

$$\mathcal{U}(\alpha) = \left\{\varepsilon : \ \int_0^T \varepsilon^T A \varepsilon \, \mathrm{d}t \leq \alpha^2\right\} \tag{6.26}$$

[3]It may be that instead of measurements we have an empirical input/output model which allows us to calculate the state and its derivatives for any class of inputs.

The upper limit, T, can be infinity. Note that the uncertainty parameters of these two models, eqs.(6.25) and (6.26), have different physical units.

6.4.3 Multi-Hypothesis Identification

The data-structure of this problem — being able to perform measurements or having an empirical input/output model — is particularly conducive to multi-hypothesis testing, which we now describe.

We demonstrate the general procedure with an example. Suppose the system is one-dimensional and that the nominal model is linear:

$$\widehat{L} = \widehat{m}\ddot{x} + \widehat{c}\dot{x} + \widehat{k}x \tag{6.27}$$

where the coefficients $\widehat{m}$, $\widehat{c}$ and $\widehat{k}$ are approximately correct. Let us assume that we know that the unknown function h is non-linear and of the following form:

$$h(x, \dot{x}) = b_1|x|x + b_2|\dot{x}|\dot{x} \tag{6.28}$$

However, we do not know the coefficients (b_1, b_2). We choose r hypothesized non-linear models, denoted by their coefficient vectors, $b^{(1)}, \ldots, b^{(r)}$. Applying any input $f(t)$ to the system yields the state and its derivatives, x, $\dot{x}$ and $\ddot{x}$. This allows us to calculate $\widehat{L}(x, \dot{x}, \ddot{x})$, and to approximate the *actual* value of h by $\widehat{L}+f$. But we can also use x, $\dot{x}$ and $\ddot{x}$ and eq.(6.28) to calculate h for each hypothesis, $b^{(1)}, \ldots, b^{(r)}$. Let us denote these *hypothesized* h values by $h^{(1)}, \ldots, h^{(r)}$. We now compare each hypothesis against the anticipated h, and choose the closest hypothesis. More precisely, we choose the cth model if:

$$\left\| h^{(c)} - \widehat{L} - f \right\|_W = \min_{1 \le n \le r} \left\| h^{(n)} - \widehat{L} - f \right\|_W \tag{6.29}$$

where $\|\cdot\|_W$ is a weighted euclidean norm, defined as:

$$\|x\|_W^2 = x^T W x \tag{6.30}$$

where W is a real, symmetric, positive definite matrix.

6.4.4 Robustness of Asymptotic Multi-Hypothesis Algorithms

In this section we calculate the robust reliability of the multi-hypothesis algorithm in the limiting case of a *very large* number of hypotheses. In the next section we extend this analysis to the more realistic situation of a finite number of hypotheses.

As stated earlier, the robust reliability of an identification algorithm is the maximum uncertainty which is consistent with acceptable performance of the algorithm. Our criterion for acceptable performance of the algorithm is that the normed difference between the identified and actual models be no greater than Δ_{cr}:

$$\left\| h^{(c)} - h \right\|_W \leq \Delta_{cr} \tag{6.31}$$

In problem 5 we will consider an integral acceptability criterion.

We are presently considering the ideal asymptotic case of *unlimited* hypotheses, so we can reasonably expect that one of these hypotheses will very closely match the calculated response.[4] The multi-hypothesis algorithm, eq.(6.29), will cause this hypothesized model to be adopted. So, the adopted model will satisfy:

$$h^{(c)} = \widehat{L} + f \tag{6.32}$$

But the righthand side of this relation is the calculational model, which may be in error by as much as ε. When this is so, the normed difference between the identified and actual models will be:

$$\left\| h^{(c)} - h \right\|_W = \|\varepsilon\|_W \tag{6.33}$$

Let us denote this error by Δ_{asym}. How large can this error be? This is controlled by the uncertainty in ε, which belongs to a convex model. We find the maximum error of the identified model, in this asymptotic case, by solving the following optimization problem:

$$\Delta^2_{asym} = \max \varepsilon^T W \varepsilon \quad \text{subject to} \quad \varepsilon \in \mathcal{U}(\alpha) \tag{6.34}$$

For the instantaneous energy-bound convex model of eq.(6.25) this becomes:

$$\Delta^2_{asym} = \max \varepsilon^T W \varepsilon \quad \text{subject to} \quad \varepsilon^T A \varepsilon \leq \alpha^2 \tag{6.35}$$

This leads to an eigenvalue problem involving the real, symmetric positive-definite matrix A in the convex model, whose solution we developed in section 4.5.2. Denote the eigenvalues of the matrix $W^{-1/2}AW^{-1/2}$ by $\mu_1 \leq \cdots \leq \mu_N$. One finds the following solution:

$$\Delta^2_{asym} = \max_{\varepsilon \in \mathcal{U}(\alpha)} \varepsilon^T W \varepsilon = \frac{\alpha^2}{\mu_1} \tag{6.36}$$

This relation shows that the maximum error in identifying the model is proportional to α, the maximum error in the nominal model, and is inversely proportional to the least eigenvalue of $W^{-1/2}AW^{-1/2}$, combining properties of the convex model and the weighting matrix of the norm. Eq.(6.36) shows that the uncertainty parameter α^2 is "normalized" by the least eigenvalue of $W^{-1/2}AW^{-1/2}$. Conversely, we could divide the inequality in the convex model of eq.(6.25) by μ_1, producing a "scaled" A matrix and "re-scaling" the uncertainty parameter in units of μ_1.

[4]There may of course be more than one such hypothesis, but we will ignore this possibility.

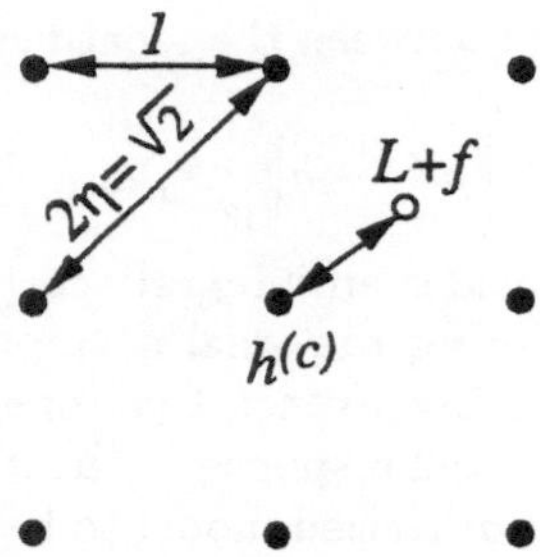

Figure 6.2: Illustration of the distance η between hypothesized responses, in a 2-dimensional search space.

Relation (6.36) can be used to express the robust reliability of this multi-hypothesis identification algorithm in the asymptotic case of unlimited hypotheses. The algorithm is acceptable if the maximum error of the model is no greater than $\Delta_{\rm cr}$, as expressed by relation (6.31). Equating the maximum error, eq.(6.36), to the critical value of the error, $\Delta_{\rm cr}$, and solving for the uncertainty parameter, α, determines the greatest allowable uncertainty, which is the robustness of the algorithm:

$$\Delta^2_{\rm asym} = \Delta^2_{\rm cr} \quad \Longrightarrow \quad \widehat{\alpha}^{\rm asym} = \Delta_{\rm cr}\sqrt{\mu_1} \tag{6.37}$$

When $\widehat{\alpha}^{\rm asym}$ is large the system is robust and can tolerate a large amount of uncertainty, implying that the multi-hypothesis identification algorithm is reliable. On the other hand, when $\widehat{\alpha}^{\rm asym}$ is small the identification is sensitive to uncertainty and the identification is unreliable.

Eq. (6.37) indicates that the robustness increases with $\Delta_{\rm cr}$: we can tolerate greater uncertainty as the performance criterion is relaxed. Also, the robustness increases with the least eigenvalue of $W^{-1/2}AW^{-1/2}$.

6.4.5 Robustness of Finite Multi-Hypothesis Algorithms

In practice, we can implement a multi-hypothesis test with at most a finite number of hypotheses, so we must modify our asymptotic analysis. In this section we calculate the robust reliability of a multi-hypothesis algorithm with a finite number of hypotheses. We can no longer expect the multi-hypothesis algorithm to supply a model which exactly matches the calculated response, unlike the asymptotic case considered in the previous section. We will suppose that our hypothesized responses are distributed approximately uniformly in a bounded domain of the search space. The distance between the best (closest) hypothesized response and the calculated response $\widehat{L}+f$ can be as large as a quantity η, for which we will develop an expression later. That is, the hypothesis chosen by the multi-hypothesis algorithm, $h^{(c)}$, can differ from the calculated response, $\widehat{L}+f$, by a grid-effect vector g, whose length can be as great as η, as shown in fig. 6.2:

$$h^{(c)} = \widehat{L} + f + g \tag{6.38}$$

But as before, in choosing the hypothesis which is closest to the nominal model $\widehat{L}$, we could also err by a vector ε, whose maximum possible norm is Δ_{asym}, which is the error of the nominal model:

$$\widehat{L} + f = L + f + \varepsilon \tag{6.39}$$

Combining the last two relations, the total error in the identified model could be as large as $\varepsilon + g$:

$$h^{(c)} = L + f + \varepsilon + g \tag{6.40}$$

So, the norm of the error could be as large as:

$$\begin{aligned}
\Delta_{\text{fin}} &= \left\| h^{(c)} - L - f \right\|_W && (6.41)\\
&= \|\varepsilon + g\|_W && (6.42)\\
&\leq \|\varepsilon\|_W + \|g\|_W && (6.43)\\
&= \Delta_{\text{asym}} + \eta && (6.44)\\
&= \frac{\alpha}{\sqrt{\mu_1}} + \eta && (6.45)
\end{aligned}$$

As in eq.(6.37), we calculate the robust reliability of the finite algorithm by equating the maximum error of the identification with the greatest acceptable error, and then solving for the uncertainty parameter:

$$\Delta_{\text{fin}} = \Delta_{\text{cr}} \quad \Longrightarrow \quad \widehat{\alpha}^{\text{fin}} = (\Delta_{\text{cr}} - \eta)\sqrt{\mu_1} \tag{6.46}$$

Comparing this with eq.(6.37), we see that the robustness of the finite multi-hypothesis algorithm is reduced by the grid effect of the finite lattice of hypotheses.

It now remains to develop an approximate expression for η, the greatest possible distance between the calculated response and the closest hypothesized response. Suppose we have r hypotheses, $h^{(1)}, \ldots, h^{(r)}$, which are vectors in $\Re^N$. We suppose these r hypotheses are more or less uniformly distributed in a volume V in N-dimensional space. Thus the volume-per-hypothesis is V/r. The equivalent hypercube with this volume has side-length $q = (V/r)^{1/N}$. We imagine the hypotheses arranged on the vertices of these equivalent hypercubes. The cube-center is the point farthest from the vertices. The distance between a vertex (where an hypothesis lies) and the cube-center (where the calculated response could lie) is $\frac{1}{2}q\sqrt{N}$. So η is estimated as:

$$\eta \approx \frac{\sqrt{N}}{2}\left(\frac{V}{r}\right)^{1/N} \tag{6.47}$$

Substituting this into eq.(6.46), the final expression for the robustness of a finite algorithm with r hypotheses in a volume V of $\Re^N$ is:

$$\widehat{\alpha}^{\text{fin}} = \left(\Delta_{\text{cr}} - \frac{\sqrt{N}}{2} \left(\frac{V}{r} \right)^{1/N} \right) \sqrt{\mu_1} \tag{6.48}$$

When $\widehat{\alpha}^{\text{fin}}$ is large the system is robust and can tolerate a large amount of uncertainty, implying that the multi-hypothesis identification algorithm is reliable. On the other hand, when $\widehat{\alpha}^{\text{fin}}$ is small the system is sensitive to uncertainty and the identification is unreliable.

Relation (6.48) shows that the robustness increases as $(V/r)^{1/N}$ with increasing number of hypotheses r. When the dimension N of the state space is large, this increase can be very slow, indicating the not-surprising need for many hypotheses. At the other extreme, when $\widehat{\alpha}^{\text{fin}}$ vanishes, the identification algorithm is extremely sensitive to the uncertainty in the nominal model; even very small error in the nominal model can result in unacceptable error in identifying the remainder, h, of the model. Equating $\widehat{\alpha}^{\text{fin}} = 0$ and solving for r results in a minimum tolerable number of hypotheses:

$$\widehat{\alpha}^{\text{fin}} = 0 \quad \Longrightarrow \quad r_{\text{min}} = \frac{V N^{N/2}}{(2\Delta_{\text{cr}})^N} \tag{6.49}$$

We can understand this expression for r_{min} as follows. Δ_{cr} is a length in $\Re^N$: the greatest acceptable distance between the chosen and actual models, $h^{(c)}$ and h. On the other hand, V is the volume in $\Re^N$ containing the hypothesized models. Let the hypercube of volume V have side-length q, so $V = q^N$. Then the minimum tolerable number of hypotheses becomes:

$$r_{\text{min}} = \left(\frac{q\sqrt{N}}{2\Delta_{\text{cr}}} \right)^N \tag{6.50}$$

This is sort of a sampling theorem. If we are searching in a region of side-length q, and seeking model-accuracy Δ_{cr}, the minimum acceptable number of hypotheses is determined by the ratio q/Δ_{cr}, and by the dimension N of the search space. Recall that r_{min} is a *bare* minimum, and that we should probably use more hypotheses. The utility of additional hypotheses is expressed by the robust reliability, eq.(6.48).

6.4.6 Hierarchical Multi-Hypothesis Algorithms

The reliability of multi-hypothesis system-identification increases as we increase the number of hypotheses which are tested, as shown in eq.(6.48). If the dimension of the search space is large, however, we will need a large number of hypotheses. Nevertheless, it may be possible to reduce the number of hypotheses (or increase the reliability) by using an hierarchical structure for the test procedure.

One qualitative scheme for hierarchical testing is as follows. With a fixed total number of hypotheses, we test a sequence of search volumes, each volume chosen on the basis of the multi-hypothesis decision of the previous stage. Each stage has a limited number of hypotheses.

Two-stage hierarchy. We begin by examining a two-stage hierarchical multi-hypothesis test procedure. We will test a total of r hypotheses, but we do not know how many hypotheses to test in the first stage. Suppose we begin by testing only $r_1 < r$ hypotheses, and subsequently choose the remaining $r_2 = r - r_1$ hypotheses on the basis of the first stage.

Let the initial search volume be V_1. From among the r_1 hypotheses tested in the first stage, the selected hypothesis is denoted $h^{(1,c)}$, and its deviation from the actual model, h, is bounded above by Δ_{fin}. Combining eqs.(6.45) and (6.47) this becomes:

$$\left\| h^{(1,c)} - h \right\|_W \le \Delta_{\text{fin}} = \frac{\alpha}{\sqrt{\mu_1}} + \frac{\sqrt{N}}{2}\left(\frac{V_1}{r_1}\right)^{1/N} \tag{6.51}$$

The volume V_2 of the search region in the second stage should be large enough to encompass the correct model. Let us choose this volume equal to a hypercube of side-length q_2, so $V_2 = q_2^N$, where we choose the side-length on the order of twice Δ_{fin} to assure that we include the actual model in the search region:

$$q_2 = \rho\left[\frac{\alpha}{\sqrt{\mu_1}} + \frac{\sqrt{N}}{2}\left(\frac{V_1}{r_1}\right)^{1/N}\right] \tag{6.52}$$

where ρ is not too different from two. If we choose $r_2 = r - r_1$ hypotheses distributed more or less uniformly in the volume V_2, and if V_2 does in fact contain the actual model, h, then the error of the estimate will again be no more than:

$$\left\| h^{(2,c)} - h \right\|_W = \Delta_{\text{fin}} \tag{6.53}$$

$$= \frac{\alpha}{\sqrt{\mu_1}} + \frac{\sqrt{N}}{2}\left(\frac{V_2}{r_2}\right)^{1/N} \tag{6.54}$$

$$= \frac{\alpha}{\sqrt{\mu_1}} + \frac{\rho\sqrt{N}}{2r_2^{1/N}}\left[\frac{\alpha}{\sqrt{\mu_1}} + \frac{\sqrt{N}}{2}\left(\frac{V_1}{r_1}\right)^{1/N}\right] \tag{6.55}$$

For our final decision we require the following bound on the error:

$$\left\| h^{(2,c)} - h \right\|_W \le \Delta_{\text{cr}} \tag{6.56}$$

Equating eq.(6.55) to Δ_{cr} and solving for the uncertainty parameter, α, we obtain the greatest uncertainty consistent with acceptable performance of the identification algorithm for this two-stage process with a total of r hypotheses,

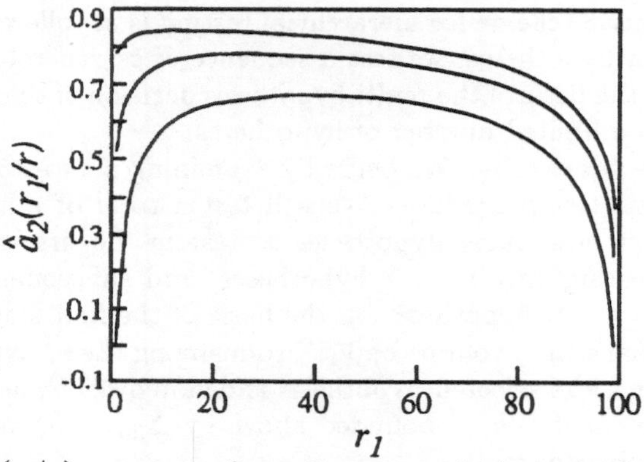

Figure 6.3: $\widehat{\alpha}_2(r_1|r)$ versus r_1 for a 2-stage multi-hypothesis identification. Parameter values: $\rho = 2$, $N = 2$, $\mu_1 = 1$, $\Delta_{\rm cr} = 1.0$. From top to bottom, $q_1 = 1, 4, 10$.

where r_1 hypotheses are used in the first stage:

$$\widehat{\alpha}_2(r_1|r) = \frac{\sqrt{\mu_1}\left(4\Delta_{\rm cr} r_1^{1/N} r_2^{1/N} - \rho N q_1\right)}{2r_1^{1/N}\left(2r_2^{1/N} + \rho\sqrt{N}\right)} \tag{6.57}$$

where $0 < r_1 < r$ and $V_1 = q_1^N$. With r fixed and with $r_2 = r - r_1$, this is readily maximized over the integer values $0 < r_1 < r$. The result is the greatest reliability for a 2-stage algorithm with r hypotheses:

$$\widehat{\alpha}_2(r) = \max_{0<r_1<r} \widehat{\alpha}_2(r_1|r) \tag{6.58}$$

We also obtain the optimal partition of the number of hypotheses in the first and second stages, $\widehat{r}_1$, and $\widehat{r}_2$.

Fig. 6.3 shows three examples of the variation of $\widehat{\alpha}_2(r_1|r)$ with r_1. The initial search volume, V_1, increases by a factor of 100 from the top to the bottom curve.

Consider first the top curve, for which the initial search volume is the least of the three curves, with $q_1 = 1$. The maximum value of the reliability occurs at $r_1 = 15$, where $\widehat{\alpha}_2(r_1|r) = 0.84$. Using eq.(6.48) to calculate the one-step reliability with the same parameter values we find $\widehat{\alpha}_1 = 0.93$, so the optimal 2-step algorithm is in fact less reliable than the one-step procedure.

We can understand this as follows. Increasing the search volume will always decrease the reliability, as seen from both eq.(6.48) and eq.(6.57), while increasing the number of hypotheses will improve the reliability. In the first stage of the two-stage process the volume is the same as in the single-stage algorithm while $r_1 = 15$ rather than $r = 100$. In the second stage both V_2 and r_2 are less than in the single-stage process. The multi-stage algorithm we have formulated, with a constant total number of hypotheses, involves a trade-off between search volume and number of hypotheses. In the present case this trade-off is less optimal than the single-stage search with the same volume and total number of hypotheses.

The initial search volume for the second curve is greater, $q_1 = 4$, and the optimal two-stage reliability is $\widehat{\alpha}_2 = 0.78$ at $r_1 = 29$, while for the one-stage identification the reliability is lower: $\widehat{\alpha}_1 = 0.72$, so now the two-stage algorithm is slightly better than the one-stage test. Finally, the lower curve, with $q_1 = 10$, has an optimal reliability of $\widehat{\alpha}_2 = 0.67$ at $r_1 = 39$. In this case, however, the one-stage identification has a reliability $\widehat{\alpha}_1 = 0.29$, which is substantially poorer than in the hierarchical implementation.

Recursive Solution for Multi-Stage Hierarchy. We now consider the general M-stage hierarchy with a total of r hypotheses and an initial search volume V_1. The task is to find the partition of the number of hypotheses per stage which maximizes the reliability. In the recursive approach we express the $(M+1)$-stage solution in terms of the M-stage solution.

The optimal partition of r hypotheses among M stages is denoted $\widehat{r}_1, \ldots, \widehat{r}_M$. This is the partition which maximizes the reliability of the identification, and this maximum reliability is denoted $\widehat{\alpha}_M(r)$. We have already solved the optimal-partition problem for $M = 2$.

Now we consider the $(M+1)$-stage process. Suppose we choose some value r_1 for the number of hypotheses in the first stage. This leaves $r - r_1$ hypotheses to be chosen in the remaining M stages. Let us denote the maximum reliability for these remaining M stages as $\widehat{\alpha}_M(r - r_1|r_1)$. But we already know how to partition a given number of hypotheses in M stages so as to maximize the reliability, and the optimal reliability for this M-stage process is $\widehat{\alpha}_M(r - r_1)$. So, having arbitrarily chosen r_1 hypotheses in the first stage of an $(M+1)$-stage process, the greatest reliability at the end is:

$$\widehat{\alpha}_{M+1}(r - r_1|r_1) = \widehat{\alpha}_M(r - r_1) \tag{6.59}$$

To find the overall optimal reliability for the $(M+1)$-stage process we must optimize $\widehat{\alpha}_{M+1}(r - r_1)$ on r_1 with r fixed:

$$\widehat{\alpha}_{M+1}(r) = \max_{r_1} \widehat{\alpha}_M(r - r_1) \tag{6.60}$$

This is a one-dimensional optimization problem, and yields the optimal partition of r hypotheses among $M+1$ stages, as well as the maximum reliability of the $(M+1)$-stage process.

6.5 Problems

1. A small mechanical device is subjected to a load which increases approximately linearly in time from zero. The load produces vibration in the device, and it is necessary that the acceleration of the device not exceed the critical value of a_{cr} for a long duration T. The vibration of the device is modelled as a one-dimensional undamped linear mass-spring system:

$$m\ddot{x}(t) + kx(t) = u(t), \quad x(0) = \dot{x}(0) = 0 \tag{6.61}$$

The equivalent mass of the model can be estimated accurately, but the stiffness is uncertain. The uncertain ramp input belongs to a slope-bound convex model:

$$\mathcal{U}(\alpha_i) = \left\{ u(t) : \ u(0) = 0, \ \left| \frac{\mathrm{d}u}{\mathrm{d}t} - s \right| \leq \alpha_i \right\} \tag{6.62}$$

For given uncertainty α_i in the load, it is necessary to decide if the device will exceed its acceleration limit for a long duration T. Is the mass-spring model reliable, and is the decision robust with respect to uncertainty of the stiffness?

2. A manufacturing process involving plastic extrusion is controlled by several temperatures and pressures, represented as positive parameters $p_1, \ldots, p_r$. These parameters are monitored, but they each fluctuate within a particular range:

$$|p_i - \overline{p}_i| \leq \rho_i \tag{6.63}$$

where $\overline{p}_i$ is the nominal value of the ith parameter and $\overline{p}_i > \rho_i$. Experience with the process has shown that if:

$$\sum_{i=1}^{r} c_i p_i \leq s_{\mathrm{cr}} \tag{6.64}$$

then the quality of the product is satisfactory. The c_i's are empirical coefficients. (a) Will the process run satisfactorily throughout the range of values of the parameters $p_1, \ldots, p_r$? (b) The performance model, eq.(6.64), is empirical, so the coefficients $c_1, \ldots, c_r$ are approximate. How robust is the decision in part (a) to uncertainty in the model coefficients $c_1, \ldots, c_r$? Suppose that each coefficient can vary by as much as a fraction $\alpha < 1$ from its nominal value:

$$\left| \frac{c_i - \overline{c}_i}{\overline{c}_i} \right| \leq \alpha \tag{6.65}$$

(c) Consider a numerical example with three parameters: two temperatures and a pressure, whose nominal values (in degrees and Pascals) are

$(\overline{p}_1, \overline{p}_2, \overline{p}_3) = (150,\ 230,\ 2.4)$. The range parameters are $(\rho_1, \rho_2, \rho_3) = (3,\ 6.9,\ 0.04)$. The nominal model coefficients are $(\overline{c}_1, \overline{c}_2, \overline{c}_3) = (0.7,\ 0.4,\ -30)$. The threshold for failure is $s_{cr} = 140$. Evaluate the robustness of the performance model for these parameter values.

3. ‡ *Modification of problem 2.* Instead of the linear performance criterion of eq.(6.64) consider the following linear-quadratic criterion:

$$p^T c + p^T D p \leq s_{cr} \tag{6.66}$$

where p is the vector of process parameters, c is a vector of empirical coefficients and D is a real symmetric matrix of empirical coefficients. (a) Will the performance be satisfactory over the range of parameter intervals specified by eq.(6.63)? (b) Consider an ellipsoid-bound uncertainty model for the process parameters:

$$\mathcal{P}(\alpha) = \left\{p:\ (p - \overline{p})^T A (p - \overline{p}) \leq \alpha^2\right\} \tag{6.67}$$

where $\overline{p}$ is the nominal parameter vector and A is a real, symmetric, positive definite matrix. Will the performance be satisfactory for all $p \in \mathcal{P}(\alpha)$?

4. Modify Eq.(6.3) to allow variable tolerance in the fractional error of the model along the length of the fin:

$$\mathcal{U}(\alpha) = \left\{\tau(v):\ \left|\frac{\tau(v)}{T_0(v)}\right| \leq \alpha(v)\right\} \tag{6.68}$$

How should one choose $\alpha(v)$?

5. Modify the results of section 6.4 for the integral energy-bound convex model, eq.(6.26), and the following integral acceptability criterion:

$$\int_0^T \left\|h^{(c)} - h\right\| \,\mathrm{d}t \leq \Delta_{cr} \tag{6.69}$$

The upper limit, T, can be infinite. (a) Show that the greatest asymptotic error of the multi-hypothesis algorithm is identical to eq.(6.36), though the α's are uncertainty parameters of different convex models. (b) Show that the robustness of the asymptotic algorithm will be the same as eq.(6.37), and the error and robustness of the finite algorithm will be identical in form to eqs.(6.45) and (6.46) respectively.

Chapter 7

Convex and Probabilistic Models of Uncertainty

The theory of probability is a prize flower in the garden of mathematics. Many of the most creative mathematicians have contributed to this theory, which is characterized by a subtle combination of intuition and analysis. The engineers acquired the theory of probability fairly recently from the scientists (who got it from aristocratic 17th century gamblers!) and have found it immensely useful. However, as we have seen in the previous chapters, probability is not the only mathematical tool with which we can quantify uncertainty. Robust reliability is derived from convex rather than probabilistic models of uncertainty.

The phenomenon of uncertainty can be defined probabilistically, in terms of the frequency or likelihood of recurrence of events. But this is not the only possible definition. Webster's says that *uncertainty* is "the quality or state of being uncertain: Doubt", and that *uncertain* means "indefinite, indeterminate, ... not reliable: Untrustworthy". The *Encyclopedia of Mathematics* [100, vol. 7, p.302] writes that "mathematical probability may serve as an estimate of the probability of an event in the ordinary everyday-life sense, i.e. it may render more precise "problematic" statements usually expressed by the words 'probably', 'possibly', 'very probably' etc." March analyzes individual and organizational decision making and writes that "Uncertainty is a limitation on understanding and intelligence." [64, p.178]. Galbraith, in discussing the design of complex industrial organizations, defines uncertainty as an information gap, as "the difference between the amount of information required to perform the task and the amount of information already possessed by the organization." [44, p.5] The point of dragging out these definitions is that "uncertainty" has diverse lexical and intuitive meanings, based on human experience, which are prior to our formulation of a mathematical model for quantifying that experience. Probabilistic models of uncertainty, as well as fuzzy models, convex models and other theories, are specific mathematical devices for representing various aspects of the phenomenon of uncertainty.

Convex models are particularly suited to the type of information often available in mechanical reliability analysis, and robust reliability is based on convex models of uncertainty, in contrast to classical probabilistic reliability. In this chapter we attempt to understand the basis of the difference between convex and probabilistic models of uncertainty. We begin in section 7.1 with a riddle which highlights the difference. In section 7.2 we compare the structure of probabilistic and convex models of uncertainty. In section 7.3 we identify some specific limitations of probability theory in technological applications. Finally, in section 7.4 we consider an example illustrative of the dangers of incautious probabilistic analysis of reliability. In summary, the proper choice of a model of uncertainty, whether probabilistic or convex, depends on the type of information which is available about the uncertainty.

7.1 Uncertainty Is Not Necessarily Probabilistic: The Three-Box Riddle

Statement of the three-box riddle: We know that a prize has been placed in one of three closed boxes, but we do not know which box. We are asked to choose a box, and if our choice is correct, we win the prize. For convenience, let us call the box we choose C. At least one of the two remaining boxes is, of course, empty. This remaining empty box we will call E. We will refer to the third box as T. Now the Master of Ceremonies, who knows both our choice and the correct box, opens E, shows us that it is empty, and gives us the option of changing our choice from C to T. The question is: do we have any rational basis for revising our choice?[1]

The argument against changing our bet, propounded by, say, Mr. Kahn, maintains that we had no basis for preferring any of the three boxes before E was opened, and we have no basis for preferring among the remaining two afterwards. We certainly should not now choose E, because we know with complete certainty that the prize is not in E. However, complete uncertainty reigns over the disposition of the prize between C and T. There is no rational basis for changing our bet, says Kahn.

In favor of changing the bet is Mr. Prow, who jumps urgently from his seat exclaiming that T is now strongly preferred over C. He explains that, before the empty box was opened there was equal probability that the prize was in any of the three boxes. This implies that the probability that it is in C is 1/3 and the probability that it is in *either* E *or* T must be 2/3. The M.C.'s choice between E and T provides information about the relative probability of E to T, but does not alter the probability that the prize is in C. After E

[1] For background on the popular discussion of this riddle see [96].

is opened and shown to be empty, we must conclude that T is the favorite over C by two to one.

To clinch the argument against Kahn, Mr. Prow begs to consider a hypothetical situation. Suppose there were 100 boxes rather than only 3. We know nothing about the location of the prize so, says Prow, the probability of it occurring in any particular box is 1/100. We choose one box, call it C, and the M.C. opens 98 empty boxes, leaving closed just C and one remaining box, call it T. Since the probability is 99/100 that the prize falls in one of the 99 boxes other than C, and all but one of these 99 boxes have been eliminated, we must certainly switch our bet from C to T.

Analyzing this dispute between Prow and Kahn we recognize that Prow has made one assumption which, if valid, vindicates his argument. Prow has translated his lack of certainty about the location of the prize into probabilistic information, by asserting that the probability is equal for each box. But this assertion is not the only probabilistic model which is consistent with uncertainty about the location of the prize. Let us imagine that the M.C. has been playing the three-box game daily for years, and that he makes a habit of putting the prize in the left-most box 80% of the time, and 10% of the time in each of the other boxes. Unless we have prior knowledge about the personal habits of the M.C. we would never suspect such behaviour, but then again, Mr. Prow is presuming different prior knowledge by adopting the equal-probability hypothesis. If the location-probabilities were (0.8, 0.1, 0.1) we would still be uncertain about the location of the prize, but we would choose the 80%-box and not change our bet when one of the other boxes is shown to be empty.

Mr. Prow's presumption is unwarranted, and his conclusion is unjustified. He has been thinking probabilistically about a problem involving uncertainty but for which no probabilistic information is available. To clarify this assertion, let us consider the infinity of probability models which could govern the M.C.'s behaviour, from which Prow has chosen just one. Let p_1 and p_2 denote the probability that the M.C. will place the prize in the 1st and 2nd box, respectively. The probability for the 3rd box is determined by p_1 and p_2 as $p_3 = 1 - p_1 - p_2$. The triangle in fig. 7.1 delimits the set of all possibilities for (p_1, p_2), and the point at $(1/3, 1/3)$ represents Prow's assumption. The initial information is that the prize is in one of the three boxes. Beyond this we know nothing, so a choice of one probability model over another would be arbitrary, not based on available data, and thus not justified, not 'rational'. Mr. Prow's assumption that $(p_1, p_2) = (1/3, 1/3)$ is unwarranted and his argument, which otherwise is correct, must be rejected.

It might be objected that the uniform distribution, $(p_1, p_2, p_3) = (1/3, 1/3, 1/3)$, is the "most uncertain" and thus justified by our complete lack of knowledge about the disposition of the prize. The fallacy in this reasoning is that, though the uniform distribution is entropically the "most uncertain" from among all probability distributions (as we will shortly explain), the

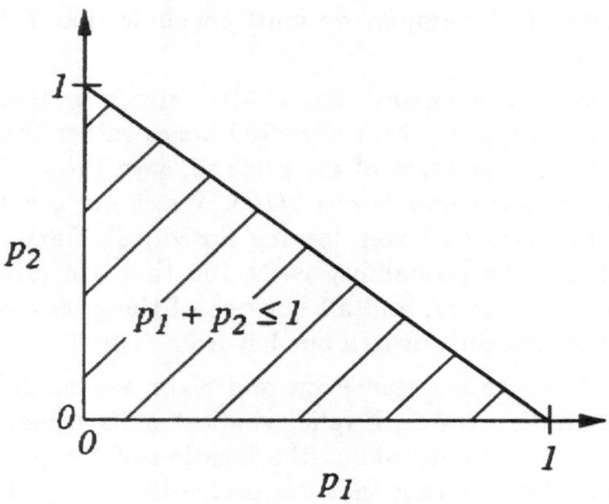

Figure 7.1: The probability triangle.

choice of this (or any other) particular distribution presumes more information than we have. To use Galbraith's expression [44], the uncertainty in this problem is a matter of the "information gap" between what we know (there is only one prize) and what we need to know (where the prize is) in order to make a decision.

Let us consider the concept of entropy. Imagine an experiment which can result in one of N possible outcomes. Let $p_1, p_2, \ldots, p_N$ denote the probabilities of these outcomes. Knowledge of these probabilities constitutes information (in the ordinary lexical sense) about the experiment. In statistical information theory one quantifies this probabilistic information with the concept of entropy of the distribution, defined as:

$$H(p_1, \ldots, p_N) = -\sum_{n=1}^{N} p_n \ln p_n \tag{7.1}$$

The entropy, H, is a non-negative number whose magnitude expresses a lack of information about the possible outcome. In the three-box game, with $N = 3$ and $p_1 = p_2 = p_3 = 1/3$ the entropy is $H \approx 1.0986$, while if $p_1 = 0.9$, $p_2 = p_3 = 0.05$ the entropy is less: $H \approx 0.3944$, corresponding to greater certainty about the outcome. If one outcome is inevitable and the others never occur (e.g. $p_1 = 1$, $p_2 = p_3 = 0$) then the entropy is zero, representing the complete certainty of the result.

Of all distributions $p_1, p_2, \ldots, p_N$, the uniform distribution $p_1 = p_2 = \cdots = p_N = 1/N$ has the greatest entropy. For instance with $N = 3$, if we bias the distribution slightly from the uniform case, say $p_1 = 0.34$ and $p_2 = p_3 = 0.33$ the entropy decreases from ~ 1.0986 for the uniform case to ~ 1.0985. The uniform distribution has maximum entropy, maximum uncertainty, *from among all possible probability distributions*. We must emphatically insist, however, that maximum entropy does not imply complete absence of information. The fact that we know a probability distribution is itself knowledge about possible outcomes. The uniform distribution is most decidedly an expression of information. If we know the distribution is uniform we will act one way, if we know it is another distribution, we may act differently. This is precisely Mr. Kahn's argument: we have *no* information, not even probabilistic, about the three boxes before opening box E, and *no* information about the two boxes afterwards. The information gap is greater than if we knew p_1, p_2, p_3. Hence, there is no rational basis for revising our choice from C to T.

This riddle suggests two conclusions which are important for our discussion. First, Mr. Prow disclosed a predilection for the uniform probability density function; we found his preference to be unwarranted. Second, the riddle hinges on uncertainty, which, as Mr. Kahn argues, must be managed without recourse to probabilistic reasoning. The uncertainty concerning the location of the prize must be analyzed without thinking probabilistically. The theory of probability is a collection of mathematical tools for analyzing uncertainty; it is not the only method. Uncertainty is not, necessarily, probabilistic.

7.2 Models of Uncertainty: A Comparison

Both probabilistic and non-probabilistic concepts of reliability, when applied to design problems, attempt to optimize the system with respect to the uncertain factors which influence it. Different though sometimes overlapping information concerning these uncertainties is required by the two concepts of reliability.

Any probabilistic theory contains two main components: sets of events, and a measure-function defined on these sets. Typically, the sets of events are quite inclusive. For example, the normal distribution extends over the entire real numbers, and probabilities are defined for all subsets. This extravagant gaussian assumption — that anything can occur — is tempered by the probability density function (pdf) which expresses the relative frequency of occurrence of different sets of events.

The convex-model concept of uncertainty is also based on sets of events, but no measure-function on these sets is defined. Instead, information about the uncertainties is invested in the structure of the event-sets. Convex models express the *clustering* of the uncertain events, while probabilistic models quantify the uncertainty in terms of the *frequency of recurrence* of events. A convex model typically contains less information than a probabilistic model of uncertainty.

A convex model is a convex set of functions or vectors, where each element of the set represents a possible realization of an uncertain phenomenon. Given specific though limited information which characterizes the uncertain events, one can define the set of *all* functions consistent with this information. This set is often convex, hence the name, convex model. In the previous chapters we have seen many examples of convex models of uncertainty. For instance, the energy-bound convex models define sets of functions consistent with a given bound on the energy, while the spectral convex model defines the set of all functions consistent with specific spectral information. Envelope-bound convex models delimit the range of variation of uncertain functions to be consistent with given information, and are a generalization, for functions, of the idea of interval-arithmetic for uncertain parameter values.

The procedure by which one formulates a convex model is basically different from the usual method for specifying a stochastic model. In stochastic formulations one often chooses the form of the model, e.g. gaussian, and then determines the coefficients of that model (mean and covariance in the gaussian case). This procedure can work quite well when the *form* of the model is correct, for then the model-parameters can usually be estimated accurately without the need to sample too extensively. This is because, as in the gaussian model, the parameters can be related to the bulk of events which hover around the mean.

On the other hand, if the *form* of the stochastic model is only approximately correct, then the tails of the calibrated stochastic model may differ substantially from the tails of the actual distribution. This is because the model-parameters, related to low-order moments, are determined from typical rather than rare events. In this case, design decisions will be satisfactory for the bulk of occurrences, but may be less than optimal for rare events. It is the rare events — catastrophes, for example — which are often of greatest concern to the designer. The sub-optimality may be manifested as either an over-conservative or an unsafe design.

We have explained that probabilistic and convex models of uncertainty are structurally different. The former involve probability densities defined on sets of events. The latter involve no measure-functions, but instead exploit available information about the uncertainty to characterize the clustering of events. However, this difference can sometimes be viewed as a matter of degree. In the example of section 7.4 we will use a convex model to define a set of allowed probability density functions. An ambient pressure is known to be uncertain, an approximate pdf for this pressure is available, and a convex model describes the set of all possible densities. One could legitimately view this hybrid probabilistic-convex model of uncertainty as simply a collection of probabilistic models. One must recognize, however, that we have no probabilistic information describing the frequency of occurrence of the different probability densities occurring in the convex model. This is more a matter of taste than substance. The crucial point is the wide range of set-theoretic possibilities for representing uncertainty without specifying probability, and the affinity of set-theoretic uncertainty models to the information (or lack thereof) in many technological contexts. Furthermore, seemingly similar probabilistic and convex models of uncertainty can have very different implications for design and reliability. However, we must first discuss the limitations of probability theory which sometimes arise in technological applications.

7.3 Limitations of Probability

The mathematical theory of probability has proven useful in many technological applications. However, it has limitations which, when clearly identified, facilitate our understanding of the non-probabilistic alternatives.[2]

One criticism of probabilistic concepts of uncertainty arises in discussion of prior probability and Bayesian inference and decision theory. A classical objection to Bayesian statistics hits at the source of the prior distribution and utility functions. As Isaac Levi asserts: "Strict Bayesians are legitimately challenged to tell us where they get their numbers." [58, p.387]. In outlining "the general statistical decision problem", Fenstad notes that "[T]he difficulty arises in connection with the prior measure .... Does every set of alternatives carry a probability?" [41, pp.2–3] and if so, what does it mean? Furthermore, uniqueness in formulating prior distributions is illusive: a given quantity of prior information is often not represented by a unique prior probability distribution [85]. This is precisely the difficulty confronted in the 3-box riddle.

Before the twentieth century, it was common to identify ignorance of likelihoods with equality of probabilities. Many of Lewis Carroll's probabilistic riddles are based upon this assumption [26]. Also, it appears in the solution of Buffon's needle-problem, from the 18th century. On Buffon's solution Coolidge has remarked that Buffon failed to recognize "the great dangers involved in assuming the equally likely" [30]. John Venn also used the uniform distribution as the fundamental device for describing lack of information. He recognized, however, that "[a]ny attempt to draw inferences from the assumption of (uniform) random arrangement must postulate the occurrence of this particular state of things at some stage or other. But there is often considerable difficulty, leading occasionally to some arbitrariness, in deciding the particular stage at which it ought to be introduced." [99, p.97]. We hear the echoes of Mr. Kahn's objection to Mr. Prow's argument based on equal prior probabilities.

The difficulty of quantifying prior knowledge is seen quite clearly in such quandries as the 3-box riddle, the prisoner's dilemma [51] and similar problems [45] where alternative decisions each seem fully consistent with the initial information. Considering the criticism of Bayesian priors, together with these riddles whose formulation is sparse and simple yet whose resolution has taxed the attention of many people, one may be inclined to agree with Kyburg that "it might be the case that some novel procedure could be used in a decision

[2] This section is based in part on [16].

theory that is based on some non-probabilistic measure of uncertainty." [55, p.189].

Perhaps such thinking as this led Suppes and Zanotti to stress the "distinction between indeterminacy and uncertainty". Their concepts of upper and lower probabilities "are defined in a purely set-theoretical way and thus do not depend ... on explicit probability considerations." [93, p.427]. They continue:

> For a strict Bayesian there is no indeterminacy, for he would postulate a prior probability ... and thereby obtain a standard random variable ... . The concept of indeterminacy is a concept for those who hold that not all sources of error, lack of certain knowledge, etc., are to be covered by a probability distribution, but may be expressed in other ways, in particular, by random relations as generalizations of random variables, and by the resulting concepts of upper and lower probabilities. [93, p.434].

In a different vein, we must mention the reductionist view, as expressed by De Finetti: "Probabilistic reasoning — always to be understood as subjective — merely stems from our being uncertain about something." Uncertainty is the more primitive concept, while probability is a mathematical construction: "probability does not exist" [32, p.x]. Indeed, in Kolmogorov's 1933 axiomatization of probability, this theory is put in its "natural place, among the general notions of modern mathematics", with no more than a passing reference to the "concrete physical problems" from which probability theory arose. [54, p.v]. This formalistic attitude might suggest the possibility of other mathematical theories describing the same phenomena yet subject to different, non-probabilistic, interpretation. (De Finetti, however, does not seem to have this in mind [31].)

Let us consider the statisticians themselves. The theories of distribution-free inference and non-parametric statistics [47, 50] are motivated by the need to draw conclusions without assuming specific probabilistic properties for the underlying populations from which data are drawn. One can not impute 'non-probabilistic' tendencies to the proponents of these statistical theories. However, the considerable interest in non-parametric statistics attests to the difficulty one may encounter in implementing, or justifying, those statistical methods which are based on assuming specific prior or conditional probability distributions.

One of the primary driving forces in the origin of non-probabilistic models of uncertainty in the engineering community has been precisely this difficulty. Referring to turbulent wind fluctuations acting on transport aircraft or tall buildings, Sobczyk and Spencer [86, p.89] enumerate numerous complicating factors and conclude that "the engineering analysis of fatigue reliability assumes some *standard* representations of the spectrum of a turbulent wind." (Italics in the original). The assumption of standard representations arises from the difficulty of verifying more specific models. Considering steel offshore platforms they assert that "the establishment of standard load spectra ... [is] much more difficult than for aircraft structures."

In a similar vein, Murota and Ikeda develop a theory for buckling of trusses with geometrical imperfections, and comment that they

> have employed random imperfections ... although it is somewhat hypothetical at this stage, since the probability distribution cannot be known precisely in practice. The present analysis is not independent of the hypothetical distribution, and the quantitative aspects of the results will have limitations in applicability. However, the qualitative aspects of the conclusions will remain valid for a wide range of probability distributions. [67].

Design-for-reliability would seem to depend on quantitative results, not only qualitative ones.

Probabilistic models have been used in recent decades to represent the uncertainty of vibrating structures [36, 88] and seismic ground motion [89, 106]. The concern about these models arises from the fact that a stochastic model represents typical events much more reliably than rare events, especially when the model is based on limited information. In discussing fatigue failure of offshore structures, Hartt and Lin comment that it is the "extreme, infrequent stress excursions which may be important either with regard to direct damage or to subsequent interaction effects." [49, p.91]. Rare events in probabilistic models are described by the tails of the distribution, while probability distributions are usually specified in terms of mean and mean-variation parameters. This makes probabilistic models risky design tools, since it is rare events, the catastrophic ones, which must underlie the reliable design.

Finally, let us consider an attitude to uncertainty arising in the realm of operations research.

> When a probabilistic description of the unknown elements is at hand, ... one is naturally led to consider stochastic models. When only partial information, or no information at all, is available, however, there is understandably a reluctance to rely on such models. In presuming that probability distributions exist they seem inherently misdirected. Besides, the problems of stochastic optimization that they lead to can be notoriously hard to solve. [79, p.119]

This expresses quite clearly the difficulty with probabilistic analysis of the 3-box riddle.

We have seen numerous suggestions, both among philosophers and technologists, that one's thinking about uncertainty can, and sometimes should, be non-probabilistic.

7.4 Sensitivity of the Failure Probability: An Example

Let us consider failure by rupture of a long cylindrical tube subject to uncertain internal pressure, P. From the perspective of probabilistic reliability, we wish to choose the wall thickness to assure that the probabilility of failure is no larger than a specified value. However, we do not precisely know the probability density function (pdf) of the pressure. We will see that small errors in the pdf can lead to large errors in the probability of failure. This section is based in part on [14]. A similar example, with a different formulation of the uncertainty, is discussed in [20, pp.11–13].

7.4.1 Uncertainty in the PDF of the Load

The equivalent stress for Tresca's maximum shear stress failure criterion is:

$$\sigma_{\rm eq} = P/\rho, \quad \text{where} \quad \rho = \left(r_2^2 - r_1^2\right)/2r_2^2 \tag{7.2}$$

where r_1 and r_2 are the inner and outer radii, respectively. Failure occurs if $\sigma_{\rm eq}$ exceeds the yield stress, $\sigma_{\rm y}$. If the wall thickness, h, is very small, $h = r_2 - r_1 \ll r_1$, then $\rho \approx h/r_1$.

We suppose that the pdf of the pressure is in fact a complicated and imprecisely known function. For design purposes, however, we approximate the pdf as an exponential density:

$$f_0(p) = \beta_0 e^{-\beta_0 p}, \quad p \geq 0 \tag{7.3}$$

In fact the real pdf is:

$$f_\eta(p) = [\xi + \eta(p)]\, e^{-\beta_0 p}, \quad p \geq 0 \tag{7.4}$$

where $\eta(p)$ is an unknown function, and ξ is a constant which normalizes the pdf:

$$\xi = \beta_0 \left[1 - \int_0^\infty \eta(p) e^{-\beta_0 p}\, {\rm d}p\right] \tag{7.5}$$

If $\eta(p)$ is constant, then f_0 and f_η are identical. If $|\eta(p)| \ll \beta_0$, then $f_0(p)$ would seem to be a good approximation to $f_\eta(p)$.

We will use an envelope-bound convex model to represent the allowed range of variation of the functions $\eta(p)$. The set of possible η-functions is:

$$\mathcal{F} = \{\eta(p) : \ \eta_1(p) \leq \eta(p) \leq \eta_2(p)\} \tag{7.6}$$

where $\eta_1(p)$ and $\eta_2(p)$ are known non-negative functions[3] which envelop the range of variation of the perturbation, $\eta(p)$.

[3] Care must be taken in the choice of η_1 and η_2 to assure that each f_η is always non-negative and normalizable by ξ.

The mean of p on $f_0(p)$ is $1/\beta_0$. Employing eq.(7.5) one finds that the mean of p on f_η is:

$$E_{f_\eta}(p) = \frac{1}{\beta_0} + \int_0^\infty \left(p - \frac{1}{\beta_0}\right) \eta(p) e^{-\beta_0 p}\, dp \tag{7.7}$$

If the perturbation, $\eta(p)$, is small, or occurs far out on the tail of the exponential, then the means of f_0 and f_η are nearly equal.

The maximum mean, for any η in $\mathcal{F}$, occurs when $\eta(p)$ switches from the lower to the upper envelope when the rest of the integrand of eq.(7.7) changes sign from negative to positive:

$$\begin{aligned} \max_{\eta \in \mathcal{F}} E_{f_\eta}(p) &= \frac{1}{\beta_0} + \int_0^{1/\beta_0} \left(p - \frac{1}{\beta_0}\right) \eta_1(p) e^{-\beta_0 p}\, dp \\ &\quad + \int_{1/\beta_0}^\infty \left(p - \frac{1}{\beta_0}\right) \eta_2(p) e^{-\beta_0 p}\, dp \end{aligned} \tag{7.8}$$

Let us choose the following envelope functions, for which one can verify that the densities $f_\eta(p)$ are non-negative and normalizable by ξ for all η-functions if ν is sufficiently small.

$$\eta_1(p) = 0 \tag{7.9}$$

$$\eta_2(p) = \begin{cases} 0 & p < p_1 \\ \nu e^{-\gamma(p-p_1)} & p \geq p_1 \end{cases} \tag{7.10}$$

The actual pdf decays exponentially like f_0 up to a pressure p_1, beyond which the form of the pdf may deviate from f_0. When ν is small and p_1 is large, $\mathcal{F}$ defines small uncertain deviations from f_0.

Assuming $p_1 \geq 1/\beta_0$, one finds the maximum expectation on f_η, for $\eta \in \mathcal{F}$, to be:

$$\max_{\eta \in \mathcal{F}} E_{f_\eta}(p) = \frac{1}{\beta_0} \left[1 + \nu \frac{p_1 \beta_0 (\gamma + \beta_0) - \gamma}{(\gamma + \beta_0)^2} e^{-\beta_0 p_1}\right] \tag{7.11}$$

For example, let $\beta_0 = \gamma = 1$, $p_1 = 4$ and $\nu = 10^{-3}$. Then $E_{f_0}(p) = 1$, while $\max_\eta E_{f_\eta}(p) - E_{f_0}(p) \approx 3.2 \times 10^{-5}$. Thus the actual pdf is exponential over 4 standard deviations, up to $p = p_1$, and the fractional error of the mean is quite small, only 3.2×10^{-5}. In this example it would be difficult to distinguish between f_0 and f_η, for any $\eta \in \mathcal{F}$. Yet we will see that the failure probabilities and design decisions can be quite different for the various probability densities allowed by $\mathcal{F}$.

7.4.2 Sensitivity of the Failure Probability

The probability of failure by yielding equals the probability that the pressure will rise to such a level that the equivalent stress will exceed the yield stress.

For pdf f_η, the probability of failure is:

$$\varphi_\eta = \mathrm{Prob}\,(\sigma_{\mathrm{eq}} \geq \sigma_{\mathrm{y}}) = \mathrm{Prob}\,(P \geq \sigma_{\mathrm{y}}\rho) = \int_{\sigma_{\mathrm{y}}\rho}^{\infty} f_\eta(p)\,\mathrm{d}p \qquad (7.12)$$

$$= e^{-\beta_0\sigma_{\mathrm{y}}\rho} + \int_0^\infty \left[H(p-\sigma_{\mathrm{y}}\rho) - e^{-\beta_0\sigma_{\mathrm{y}}\rho}\right]\eta(p)e^{-\beta_0 p}\,\mathrm{d}p \qquad (7.13)$$

where ξ from eq.(7.5) has been substituted into eq.(7.4), and where $H(x) = 1$ if $x \geq 0$, and $H(x) = 0$ otherwise. The first term in eq.(7.13) is the probability of failure based on the nominal pdf of the pressure, f_0; the second term expresses the contribution of the uncertainty in $f_\eta(p)$.

It is an elementary matter to evaluate the greatest probability of failure, for any η-function in $\mathcal{F}$. The maximum of φ_η in eq.(7.13) occurs when $\eta(p)$ switches from its lower to its upper envelope as the term in square brackets changes in sign from negative to positive. Assume that $\sigma_{\mathrm{y}}\rho \leq p_1$. One finds the maximum probability of failure, with the envelopes of eqs.(7.9) and (7.10), to be:

$$\widehat{\varphi} = \max_{\eta\in\mathcal{F}} \varphi_\eta \qquad (7.14)$$

$$= e^{-\beta_0\sigma_{\mathrm{y}}\rho} + \frac{\nu\left(1-e^{-\beta_0\sigma_{\mathrm{y}}\rho}\right)e^{-\beta_0 p_1}}{\beta_0+\gamma} \qquad (7.15)$$

Let us consider the thin-walled case, so $\rho \approx h/r_1$. From the strict probabilistic point of view, it is reasonable to choose the tube-wall thickness, h, on the basis of the available probabilistic information, $f_0(p)$. One chooses h to achieve a specified probability of failure, φ_0, from eq.(7.13) with $\eta = 0$:

$$\varphi_0 = e^{-\beta_0\sigma_{\mathrm{y}}h/r_1} \qquad (7.16)$$

Combining eqs.(7.15) and (7.16), we can relate the maximum probability of failure, $\widehat{\varphi}$, to the nominal probability of failure, φ_0, as:

$$\widehat{\varphi} = \varphi_0 + \frac{\nu(1-\varphi_0)\,e^{-\beta_0 p_1}}{\beta_0+\gamma} \qquad (7.17)$$

The maximum probability of failure, $\widehat{\varphi}$, can be substantially greater than the value, φ_0, upon which the wall-thickness is chosen in the nominal probabilistic analysis. For example, choose $1 = \beta_0 = \gamma$ and $\nu = 10^{-3}$ as before. If the desired probability of failure is $\varphi_0 = 10^{-6}$ and the disturbance in $f(p)$ appears at four standard deviations from the origin ($p_1 = 4$), then eq.(7.17) indicates that $\widehat{\varphi}/\varphi_0 = 10.2$. That is, the actual probability of error could be as much as 10 times the design value. $\widehat{\varphi}$ is of course still a small number, but not as small as φ_0.

7.4.3 Design Implications

Let us continue with the thin-walled case. In the nominal probabilistic analysis the wall-thickness is chosen by inverting eq.(7.16) as:

$$h_{\rm p} = -\frac{r_1}{\beta_0 \sigma_{\rm y}} \ln \varphi_0 \tag{7.18}$$

If we include the non-probabilistic information about the uncertainty in the pdf, namely the convex model $\mathcal{F}$, then the wall thickness is chosen by equating $\widehat{\varphi}$ to φ_0 and inverting eq.(7.15):

$$h_{\rm cm} = -\frac{r_1}{\beta_0 \sigma_{\rm y}} \ln \frac{\varphi_0 - \zeta}{1 - \zeta} \tag{7.19}$$

where $\zeta = \nu \exp(-\beta_0 p_1)/(\beta_0 + \gamma)$. (Note that eq.(7.15) cannot be solved for h unless $\zeta < \varphi_0$. However, this is an artifact of the thin-wall restriction.)

It is now possible to compare the strict probabilistic design, $h_{\rm p}$, with the probabilistic design which has been augmented by a convex model for uncertainty in the pdf of the pressure, $h_{\rm cm}$. For example, suppose $1 = \beta_0 = \gamma$, $\nu = 10^{-3}$ and $p_1 = 4$ as before. Then $\zeta = 9.158 \times 10^{-6}$, and (augmented) thin-walled designs are available at any reliability $\varphi_0 > \zeta$. For example, a probability of failure of $\varphi_0 = 10^{-5}$ results in a ratio of the design thicknesses of $h_{\rm cm}/h_{\rm p} = 1.2$; the ordinary probabilistic design is 20% too thin.

The strict probabilistic design is under-conservative in this example because the actual pdf functions, $f_\eta(p)$, are all biased (very slightly) towards higher pressures than the nominal pdf, $f_0(p)$. If the functions $\eta(p)$ were slightly negative rather than slightly positive the reverse situation would arise: the strict probabilistic design would be overly conservative.

The point of this example is that very small uncertainties in the pdf, located far from the bulk of events, are difficult to detect but cause substantial inaccuracy in both design-decisions and assessment of failure-probability.

7.5 Problems

1. Following is a riddle posed and "solved" by Lewis Carroll [26, riddle 72]. Is his argument correct? If not, where are his errors?

   > "A bag contains 2 counters, as to which nothing is known except that each is either black or white. Ascertain their colours without taking them out of the bag."

   Answer: "One is black, and the other white."

   Solution:

"We know that, if a bag contained 3 counters, 2 being black and one white, the chance of drawing a black one would be 2/3— and that any *other* state of things would *not* give this chance.

"Now the chances, that the given bag contains (α) BB, (β) BW, (γ) WW, are respectively 1/4, 1/2, 1/4.

"Add a black counter.

"Then the chances, that it contains (α) BBB, (β) BWB, (γ) WWB, are, as before, 1/4, 1/2, 1/4.

"Hence the chance, of now drawing a black one,

$$= \frac{1}{4} \cdot 1 + \frac{1}{2} \cdot \frac{2}{3} + \frac{1}{4} \cdot \frac{1}{3} = \frac{2}{3}$$

"Hence the bag now contains BBW (since any *other* state of things would *not* give this chance.

"Hence, before the black counter was added, it contained BW, i.e. one black counter and one white."

2. *Modification of the 3-box problem.* Suppose that we have imprecise probabilistic information about the location of the prize. Let $\overline{p}_i$ be the nominal probability that the prize is in the ith box, where $\overline{p}_1 + \overline{p}_2 + \overline{p}_3 = 1$. However, these values are uncertain, and are constrained to the set:

$$\mathcal{P}(\alpha) = \left\{ (p_1, p_2) : \ (p_1 - \overline{p}_1)^2 + (p_2 - \overline{p}_2)^2 \leq \alpha^2 \right\} \tag{7.20}$$

Given values $(\overline{p}_1, \overline{p}_2)$, we will choose the box whose probability is higher. (a) Evaluate the robustness of this decision with respect to the uncertainty α. That is, what is the greatest value of $\widehat{\alpha}$ such that the decision is the same for any doublet $(p_1, p_2) \in \mathcal{P}(\alpha)$, if $\alpha < \widehat{\alpha}$? In other words, how much uncertainty can the decision algorithm tolerate without altering the decision? (b) The robustness is zero when $\overline{p}_1 = \overline{p}_2 = \overline{p}_3 = 1/3$. Discuss the implications of this for the original 3-box problem.

Chapter 8

Robust Reliability and the Poisson Process

Throughout the first 6 chapters we have exclusively used convex models of uncertainty. The primary practical motivation for avoiding the use of probabilistic models is the frequent lack of sufficient information to verify the choice of the probabilistic model. In chapter 7 we demonstrated that even small inaccuracies in the probability density can have far reaching repercussions on the reliability analysis.

In some situations, however, one does have information with which to select a probabilistic model. In particular, events distributed in time or space as discrete occurrences can sometimes be described by the Poisson process or one of its extensions, whose basic assumption may be tested against fragmentary prior knowledge about the process. In this chapter we discuss the exploitation of this sort of probabilistic information in conjunction with robust reliability analysis based on convex models. That is, we will consider a hybridization of probabilistic and robust reliability. We assume that the reader is familiar with the basic concepts of probability.

In section 8.1 we derive and discuss the Poisson distribution. In section 8.2 we develop a hybrid robust-probabilistic reliability analysis of a dynamic system with uncertain loads. In section 8.3 we do the same for a thin-walled shell with uncertain geometrical imperfections. Finally, in section 8.4 we combine probabilistic and convex uncertainty information in studying fatigue reliability.

8.1 The Poisson Distribution

The Poisson process attempts to describe the probability distribution of discrete events scattered along the time axis or distributed in space. For instance, the events can be occasional large load transients on a turbine occurring at discrete points in time, or material imperfections occurring at various points on the surface of a shell.

The Poisson process is based on a single very simple hypothesis. Considering events occurring in a random time sequence, the assumption underlying the Poisson model is that the probability of an event in an infinitesimal duration dt is independent of past events and equals $\lambda\, dt$, where λ is a constant. If the events are distributed in space rather than time we replace dt by a infinitesimal volume dv.

Obviously, this hypothesis need not always be true. It can happen that an event at one instant influences the probability of subsequent occurrences, thus violating the assumption of independence. Or the coefficient λ may vary in time or space. However, one can imagine situations in which the Poisson hypothesis holds true. Furthermore, by combining the Poisson model with convex models we will be able to combine the power of the Poisson approach with the advantages of set-models in describing complex uncertainties.

The Poisson probability of exactly n events in a duration t is denoted $P_n(t)$, where t is a continuous non-negative variable and $n = 0, 1, 2, \ldots$. The form of the functions $P_n(t)$ can be derived as follows. n events can accumulate in a duration $t + dt$ in any one of the following mutually exclusive pathways:

1. n events occurred up to time t and no event occurred during the infinitesimal interval $(t, t + dt)$.

2. $n-1$ events occurred up to time t and exactly one event occurred during $(t, t + dt)$.

3. $n - 2$ events occurred up to time t and exactly two events occurred during $(t, t + dt)$.

and so on.

The probability of an event during $(t, t + dt)$ is, by the basic Poisson assumption, λdt. Consequently, the probability of an event *not* occurring during $(t, t + dt)$ is $1 - \lambda dt$. Furthermore, the probability of two events occurring during $(t, t + dt)$ is $(\lambda dt)^2$, which we see is much smaller than λdt since dt is an infinitesimal.

Putting this all together, we obtain a probability-balance equation for the Poisson distribution [40]:

$$P_n(t + dt) = P_n(t)(1 - \lambda dt) + P_{n-1}(t)\lambda dt + \mathcal{O}(dt^2) + \cdots, \quad n = 0, 1, 2, \ldots \tag{8.1}$$

The lefthand side is the probability of n events in the duration $t+dt$. The first term on the right corresponds to pathway 1, the second term to pathway 2, and the third term to pathway 3. Dividing by dt, taking the limit $dt \to 0$, and re-arranging we obtain the following set of differential equations:

$$\frac{dP_n(t)}{dt} = \lambda[P_{n-1}(t) - P_n(t)], \quad n = 0, 1, 2, \ldots \tag{8.2}$$

We define $P_{-1} = 0$.

Usually, the initial number of events is zero, so the initial condition for (8.2) is expressed with the Kronecker delta function as:

$$P_n(0) = \delta_{n0} \tag{8.3}$$

One can prove by induction that the solution of eqs.(8.2) is:

$$P_n(t) = \frac{(\lambda t)^n \mathrm{e}^{-\lambda t}}{n!}, \quad n = 0, 1, 2, \ldots \tag{8.4}$$

This is the Poisson distribution.

We now consider a simple example illustrating the use of the Poisson distribution.

Example 1 Discrete load cycles are applied to a structure which accumulates an increment δ of damage after each cycle. Failure occurs if the cumulative damage level exceeds Δ_{cr}. The load cycles are distributed in time as a Poisson process. We will determine the probability of failure up to time t, and the average and standard deviation of the damage level at time t. Then we will evaluate the probability of failure in an infinitesimal interval $(t, t + \mathrm{d}t)$. With this result we can determine the mean and variance of the time to failure.

(a) Failure occurs when the number of cycles exceeds $N_F = \Delta_{\mathrm{cr}}/\delta$. The number of cycles is distributed in time as a Poisson variable, so the probability of n load cycles occurring in duration t is $P_n(t)$, eq.(8.4). Consequently, the probability of failure in duration t is the probability of at least N_F cycles:

$$P_{\mathrm{f}}(t) = \sum_{n \geq N_F} P_n(t) \tag{8.5}$$

$$= \sum_{n \geq N_F} \frac{(\lambda t)^n \mathrm{e}^{-\lambda t}}{n!} \tag{8.6}$$

(b) The damage is a function of the number of cycles which have occurred:

$$\Delta(t) = n(t)\delta \tag{8.7}$$

Thus the mean damage in time t is:

$$\mathrm{E}[\Delta(t)] = \mathrm{E}[n(t)]\delta = \lambda t \delta \tag{8.8}$$

where $\mathrm{E}(\cdot)$ is the expectation operator.

The variance of the damage is evaluated as follows:

$$\sigma^2(\Delta(t)) = \mathrm{E}[\Delta^2] - \mathrm{E}[\Delta]^2 \tag{8.9}$$

$$= \delta^2 \mathrm{E}[n^2] - \delta^2 \mathrm{E}(n)^2 \tag{8.10}$$

$$= \delta^2 \lambda t \tag{8.11}$$

(c) Failure occurs in the infinitesimal interval $(t, t + \mathrm{d}t)$ if the critical damage level is reached for the first time in that interval. The probability of this is the probability of reaching one less than the critical number of cycles sometime in the duration t, and then reaching N_F during $(t, t + \mathrm{d}t)$:

$$f(t)\,\mathrm{d}t = [P_{N_F-1}(t)] \times [\lambda\,\mathrm{d}t] \tag{8.12}$$

$$= \lambda \frac{(\lambda t)^{N_F-1}\mathrm{e}^{-\lambda t}}{(N_F-1)!}\,\mathrm{d}t \tag{8.13}$$

(d) We can calculate the moments of the time to failure as the moments of $f(t)$. The mean time to failure is:

$$\mathrm{E}(t) = \int_0^\infty t f(t)\,\mathrm{d}t \tag{8.14}$$

$$= \frac{\lambda^{N_F}}{(N_F-1)!}\int_0^\infty t^{N_F}\mathrm{e}^{-\lambda t}\,\mathrm{d}t \tag{8.15}$$

$$= \frac{N_F}{\lambda} \tag{8.16}$$

To calculate the variance we need the second moment:

$$\mathrm{E}(t^2) = \int_0^\infty t^2 f(t)\,\mathrm{d}t \tag{8.17}$$

$$= \frac{\lambda^{N_F}}{(N_F-1)!}\int_0^\infty t^{N_F+1}\mathrm{e}^{-\lambda t}\,\mathrm{d}t \tag{8.18}$$

$$= \frac{(N_F+1)N_F}{\lambda^2} \tag{8.19}$$

So the variance of the time to failure is:

$$\sigma_t^2 = \mathrm{E}(t^2) - \mathrm{E}(t)^2 = \frac{N_F}{\lambda^2} \tag{8.20}$$

■

8.2 Dynamic System with Uncertain Loads

Let us consider a dynamic system subject to recurring unknown time-varying loads. Each load cycle is of duration T, and the uncertainty in the waveform of the load is represented by a convex model. The recurrence of the load cycles is distributed randomly in time as a Poisson process. Each load cycle deposits a variable amount of energy in the system, depending on the waveform of the input. The system fails when the cumulative deposited energy exceeds a threshold, E_{cr}. We will evaluate the robust reliability of this system, using both the convex and the probabilistic information about the uncertainties.

The system is described as a sub-critically damped one-dimensional oscillator:

$$m\ddot{x}(t) + c\dot{x}(t) + kx(t) = u(t), \quad x(0) = \dot{x}(0) = 0 \tag{8.21}$$

The response at time t is given by eq.(4.103) (p.87). The energy dissipated by the system in a duty cycle due to the damping term, $c\dot{x}$, is expressed by eq.(4.104) (p.87). For the initial conditions we have chosen, $\dot{x}$ can be expressed as:

$$\dot{x}(t) = \int_0^t u(\tau)\frac{\mathrm{d}G(t-\tau)}{\mathrm{d}t}\,\mathrm{d}\tau \tag{8.22}$$

where $G(t-\tau)$ is defined in eq.(4.110) (p.88). Combining these relations, the energy absorbed in one duty cycle is:

$$E = c\int_0^T\left(\int_0^t u(\tau)\frac{\mathrm{d}G(t-\tau)}{\mathrm{d}t}\,\mathrm{d}\tau\right)^2\,\mathrm{d}t \tag{8.23}$$

We will represent the input as a truncated Fourier series:

$$u(t) = \sum_{n=1}^{N}\beta_n\cos\frac{k_n\pi t}{T} \tag{8.24}$$

$$= \beta^T\gamma(t) \tag{8.25}$$

where the k_n are integers, $\gamma(t)$ is a vector of known cosine functions and β is a vector of unknown Fourier coefficients. We adopt an ellipsoid-bound model for the uncertainty in β:

$$\mathcal{U}(\alpha) = \{\beta:\ \beta^T W\beta \le \alpha^2\} \tag{8.26}$$

W is a postive definite real symmetric matrix.

The energy deposited in the damped vibrating system, eq.(8.23), can be expressed as:

$$E = c\beta^T\underbrace{\int_0^T\left(\int_0^t\gamma(\tau)\frac{\mathrm{d}G(t-\tau)}{\mathrm{d}t}\,\mathrm{d}\tau\right)\left(\int_0^t\gamma^T(\theta)\frac{\mathrm{d}G(t-\theta)}{\mathrm{d}t}\,\mathrm{d}\theta\right)\mathrm{d}t}_{Z}\,\beta \tag{8.27}$$

$$= c\beta^T Z\beta \tag{8.28}$$

where Z is a real symmetric matrix which can be evaluated from prior information.

The greatest amount of energy which can be deposited in a single duty cycle is the solution of the following optimization problem:

$$E_{\max} = \max c\beta^T Z\beta \quad \text{subject to} \quad \beta\in\mathcal{U}(\alpha) \tag{8.29}$$

Using the Lagrange optimization method, one finds the following solution:

$$E_{\max} = c\alpha^2 \max \text{eig} \left[W^{-1/2} Z W^{-1/2} \right] \quad (8.30)$$

Let $\nu_{\max}$ denote this maximum eigenvalue.

The system fails when the accumulated energy absorbed from a sequence of load cycles exceeds the critical value E_{cr}. The load cycles are distributed randomly in time according to the Poisson process, but the amount of energy dissipated per load cycle depends on the input function, $u(t)$, which varies over the convex model $\mathcal{U}(\alpha)$. We have no probabilistic information about $u(t)$ or about the amount of energy E dissipated per cycle, which varies on an interval:

$$0 \le E \le c\alpha^2 \nu_{\max} \quad (8.31)$$

In other words, we do not have enough information to calculate the probability of failure in a given duration.

However, we can calculate a hybrid robust-probabilistic reliability.

In order for failure in n cycles to be possible, the load uncertainty must be large enough so that, given maximal energy dissipation in each load cycle, the cumulative energy exceeds the critical value:

$$nc\alpha^2 \nu_{\max} \ge E_{\text{cr}} \quad (8.32)$$

Solving this for α we find the critical value of α for failure in n cycles:

$$\alpha_n = \sqrt{\frac{E_{\text{cr}}}{c\nu_{\max} n}} \quad (8.33)$$

We require the system to survive at least as long as Θ. Using the Poisson model, we can calculate the probability, $P_n(\Theta)$, that n cycles will occur in the duration Θ. Failure can occur in one or more cycles, but failure cannot occur in zero cycles. So, we "randomize" α_n on the Poisson distribution, re-normalized to $n = 1, 2, \ldots$. That is, we calculate the average robustness for lifetime Θ:

$$\widehat{\alpha} = \frac{1}{1 - P_0(\Theta)} \sum_{n=1}^{\infty} \alpha_n P_n(\Theta) \quad (8.34)$$

$$= \frac{1}{1 - \mathrm{e}^{-\lambda\Theta}} \sqrt{\frac{E_{\text{cr}}}{c\nu_{\max}}} \sum_{n=1}^{\infty} \frac{\mathrm{e}^{-\lambda\Theta} (\lambda\Theta)^n}{n! \sqrt{n}} \quad (8.35)$$

α_n is the robust reliability for failure in n load cycles, and $\widehat{\alpha}$ is the average of the α_n's weighted by the Poisson probability that n cycles occur during the lifetime Θ. When $\widehat{\alpha}$ is large the system can tolerate a great amount of uncertainty in the load functions without failing, while a small value of $\widehat{\alpha}$ indicates fragility to variation of the loads. $\widehat{\alpha}$ is therefore a robust reliability, though it also incorporates a modicum of probabilistic information: the Poisson distribution in time of the load cycles.

8.3 Shells With Geometric Imperfections

A thin-walled cylindrical shell is quite sensitive to geometrical imperfections. Even small defects can lead to substantial reduction in the buckling load. We have studied the robust reliability of shells in sections 3.5 and 4.5. In this section we will combine probabilistic and convex models of uncertainty in a hybrid reliability analysis.

One can think of the geometric imperfections as small point defects scattered randomly over the surface of the shell. The Poisson process provides a simple but sometimes plausible probabilistic description of the spatial distribution of defects, whose validity is fairly easy to test.

Adopting the Poisson model, let $P_n(A)$ be the probability of exactly n defects in an area A on the surface of the shell, given by eq.(8.4) with t replaced by A.

For the moment we suppose that all the defects are identical in shape, each one covering an area A_{cr} and of amplitude δ. We suppose that the shell will fail if the total imperfection amplitude at any point exceeds the critical value Δ_{cr}. In other words, the shell will fail if the number of defects whose centers fall in an area A_{cr} exceeds the critical value:

$$N_{cr} = \frac{\Delta_{cr}}{\delta} \tag{8.36}$$

The probability of failure is the sum of probabilities that the number of defects in an area A_{cr} is N_{cr} or more:

$$P_f = \sum_{n \geq N_{cr}} P_n(A_{cr}) \tag{8.37}$$

This analysis is valid as far as it goes, if the Poisson model has been verified. However, it ignores the uncertainty in the shape of the imperfections. These dents, scratches and bends are generally inordinately complicated, and probabilistic information about their shape-uncertainty is expensive and scarce. It is natural to quantify the uncertainty in the shape of the defects with a convex model, as we have done in earlier chapters. We will then no longer be able to calculate a completely probabilistic reliability, but we will develop a hybrid robust-probabilistic measure.

Let $\eta(r)$ represent the profile of imperfection-amplitude on the shell surface. We will use a uniform-bound convex model to represent the uncertainty in the imperfection functions $\eta(r)$, in terms of a radial tolerance:

$$\mathcal{H}(\alpha) = \{\eta(r) : \ |\eta(r)| \leq \alpha\} \tag{8.38}$$

$\mathcal{H}(\alpha)$ contains imperfections whose amplitude varies in an unknown manner on the shell surface, but whose maximum amplitude is never greater than the uncertainty parameter α.

Suppose there are n defects centered in an area A_{cr}, and that their imperfection profiles are $\eta_1(r), \ldots, \eta_n(r)$. Continuing with the failure criterion underlying eq.(8.36), failure can occur with n spatially uncertain imperfections if:

$$\max_x \sum_{i=1}^{n} \eta_i(x) \geq \Delta_{cr} \tag{8.39}$$

The convex model allows the maximum on the left to be as large as $n\alpha$, so failure can occur with n defects if the uncertainty parameter is at least as large as:

$$\alpha_n = \frac{\Delta_{cr}}{n} \tag{8.40}$$

We are unable to calculate the probability of failure, since we have no probabilistic information about the imperfection profiles $\eta(r)$. However, α_n is the maximum amplitude-uncertainty which can be tolerated if n defects are present in an area A_{cr}. Likewise, $P_n(A_{cr})$ is the probability that n defects will fall in an area A_{cr}. So, like in eq.(8.34), we can randomize α_n on the re-normalized Poisson distribution:

$$\widehat{\alpha} = \frac{1}{1 - P_0(A_{cr})} \sum_{n=1}^{\infty} \alpha_n P_n(A_{cr}) \tag{8.41}$$

$$= \frac{\Delta_{cr}}{1 - e^{-\lambda A_{cr}}} \sum_{n=1}^{\infty} \frac{e^{-\lambda A_{cr}} (\lambda A_{cr})^n}{n!\, n} \tag{8.42}$$

8.4 Damage and Annealing Processes

The Poisson process is the prototype of a wide class of stochastic models known as birth and death processes [40]. In the Poisson model, random events accumulate independently, governed by the probability $\lambda\, dt$ of an occurrence in any infinitesimal interval dt. An immediate generalization is to consider two different types of events, also occurring independently, one causing accumulation and the other annihiliation of the inventory of events. This type of model is attractive for describing fatigue processes in which both damage and annealing occur. Combining this with a convex model for the uncertainty of the load functions, we will develop a hybrid robust-probabilistic analysis of fatigue reliability.

8.4.1 Birth and Death Process

The birth and death process was originally developed by Galton in the 19th century is calculate the probability of disappearance of family names [40]. It has been used to model diffusion, ecological processes, neutron dynamics in nuclear reactors [103] and many other areas. We will formulate the birth and death process as a simple representation of damage evolution with repair.

We will study the reliability in section 8.4.2, and again in section 8.4.3 with an extended birth and death model.

Let n represent the net state of damage, as the difference between the number of damage increments and the number of annealing increments: damage-minus-annealing. We will suppose that the extent of both damage and of annealing is unlimited, so that $n = \ldots - 2, -1, 0, 1, 2, \ldots$.

Let $\lambda\, dt$ be the probability of an increment of damage in the infinitesimal duration dt and let $\mu\, dt$ be the probability of an increment of annealing in dt. Assume that all events, both damage and annealing, are independent. $P_n(t)$ is the probability of damage state n at time t. In analogy to eq.(8.1), we can write the following probability-balance equation:

$$P_n(t+\mathrm{d}t) = P_n(t)(1-\lambda\,\mathrm{d}t - \mu\,\mathrm{d}t) + P_{n-1}(t)\lambda\,\mathrm{d}t + P_{n+1}(t)\mu\,\mathrm{d}t + \mathcal{O}(\mathrm{d}t^2) + \cdots \tag{8.43}$$

for $n = \ldots - 2, -1, 0, 1, 2, \ldots$. This balance of probability states that the damage state can reach the value n in a duration $t + \mathrm{d}t$ by any one of several pathways: (1) reaching n up to time t with no change in the subsequent $\mathrm{d}t$; (2) reaching $n-1$ up to time t with an additional increment of damage during the final $\mathrm{d}t$; (3) exceeding n by 1 at t followed by a single increment of annealing during $\mathrm{d}t$; (4) other pathways involving more than one change during the final infinitesimal increment of time. This is called a "birth and death process", since λ and μ represent, respectively, the "birth" and "death" of an increment of damage.

Eq.(8.43) results in the following differential equation by re-arranging and taking the limit $\mathrm{d}t \to 0$:

$$\frac{dP_n(t)}{\mathrm{d}t} = -(\lambda+\mu)P_n(t) + \lambda P_{n-1}(t) + \mu P_{n+1}(t), \quad n = \ldots -2, -1, 0, 1, 2, \ldots \tag{8.44}$$

If the initial state of damage is 0, then the initial condition is:

$$P_n(0) = \delta_{n0} \tag{8.45}$$

The mean and variance of the damage state, at time t, are:

$$\mathrm{E}[n(t)] = (\lambda - \mu)t \tag{8.46}$$

$$\sigma_n^2(t) = (\lambda + \mu)t \tag{8.47}$$

Eq.(8.46) indicates that the average damage progresses like the mean of a Poisson process whose rate parameter is $\lambda - \mu$. That is, the mean damage state expresses the competition between damage and annealing. On the other hand, the variance of the damage grows like that of a Poisson process whose rate parameter is $\lambda + \mu$, indicating that uncertainty in the damage state increases in time due to uncertainty in both the damage process and the annealing process.

Using the method of generating functions one can derive the probability distribution. For the initial conditions of eq.(8.45) one finds:

$$P_n(t) = \begin{cases} e^{-(\lambda+\mu)t} \displaystyle\sum_{k=n}^{\infty} \frac{(\lambda t)^k}{k!} \frac{(\mu t)^{k-n}}{(k-n)!} & n \geq 0 \\ e^{-(\lambda+\mu)t} \displaystyle\sum_{k=-n}^{\infty} \frac{(\mu t)^k}{k!} \frac{(\lambda t)^{k+n}}{(k+n)!} & n < 0 \end{cases} \tag{8.48}$$

8.4.2 Damage and Annealing: I

The birth and death process must be extended in order to develop a plausible representation of damage and annealing in fatigue failure. Nonetheless it does provide a rough first approximation, as we will now see.

Consider a one-dimensional damped vibrating system, eq.(8.21), driven by uncertain inputs from the Fourier ellipsoid-bound convex model, eq.(8.26). Inputs can have either of two impacts on the system: damage or annealing. For instance, small loads may result in micro-cracking, while large loads cause repair due to local plastic deformation. We have no probabilistic information about the input functions $u(t)$ except that $\lambda\, dt$ and $\mu\, dt$ are the probabilities of damage and annealing, respectively, and that the distribution in time of the damage and annealing events is described by the birth and death process.

We measure the damage and annealing in units of energy. The damage increments resulting from a single load cycle are uncertain, and vary between zero and the maximum value of $E_{\max}$, eq.(8.30). On the other hand, we suppose that the annealing increments are all of maximal magnitude and negative sign to indicate their annealing effect: $-E_{\max}$.

Failure occurs when the total net damage level exceeds $E_{\rm cr}$. This is possible with an excess n of damaging over annealing events, if the uncertainty parameter satisfies eq.(8.32). Consequently, the critical value of α for failure at net damage level n is again given precisely by eq.(8.33). Randomizing on the positive values of n we obtain the hybrid probabilistic-robust reliability for lifetime Θ as:

$$\widehat{\alpha} = \frac{\displaystyle\sum_{n=1}^{\infty} \alpha_n P_n(\Theta)}{\displaystyle\sum_{n=1}^{\infty} P_n(\Theta)} \tag{8.49}$$

where the distribution P_n is given by eq.(8.48), the solution of the birth and death process.

8.4.3 Damage and Annealing: II

We now consider a modification of the basic birth and death process, which generates a somewhat more realistic representation of damage and annealing.

Consider the evolution of damage in a dynamic system to which two types of events are occurring in parallel. One type of event causes damage, while the other has an annealing effect which ameliorates the damage.

Let $\lambda\,dt$ and $\mu\,dt$ be the probabilities for these events to occur in an infinitesimal duration dt. Let ℓ and m represent the number of occurrences of these two types of events, respectively, and denote the probability of damage state (ℓ, m) at time t by $P_{\ell m}(t)$.

We can derive a differential equation for $P_{\ell m}(t)$ following an argument similar to the derivation of eq.(8.44). The probability-balance equation for this process is:

$$P_{\ell m}(t+\mathrm{d}t) = \tag{8.50}$$
$$P_{\ell m}(t)[1-\lambda \mathrm{d}t - \mu \mathrm{d}t] + P_{\ell-1,m}(t)\lambda \mathrm{d}t + P_{\ell,m-1}(t)\mu \mathrm{d}t + \mathcal{O}(\mathrm{d}t^2) + \cdots$$

This results in the following differential equation:

$$\frac{\mathrm{d}P_{\ell m}(t)}{\mathrm{d}t} = -[\lambda+\mu]P_{\ell m}(t) + \lambda P_{\ell-1,m}(t) + \mu P_{\ell,m-1}(t) \tag{8.51}$$

for $\ell = 0, 1, 2, \ldots$ and $m = 0, 1, 2, \ldots$. We will consider the following initial conditions:

$$P_{\ell m}(0) = \delta_{\ell 0}\delta_{m0} \tag{8.52}$$

for which the solution of eq.(8.51) is:

$$P_{\ell m}(t) = \mathrm{e}^{-(\lambda+\mu)t}\,\frac{(\lambda t)^{\ell}}{\ell!}\,\frac{(\mu t)^m}{m!} \tag{8.53}$$

We note that this is simply the product of two Poisson distributions, eq.(8.4).

Again consider the damped one-dimensional vibrator, eq.(8.21), driven by inputs from either of two convex models, one representing damage events and the other annealing events. Both uncertainty models are Fourier ellipsoid-bound sets. The uncertainty in the Fourier coefficient vector of the damaging inputs is represented by:

$$\mathcal{U}_{\lambda}(\alpha_{\lambda}) = \left\{\beta:\ \beta^T W_{\lambda}\beta \le \alpha_{\lambda}^2\right\} \tag{8.54}$$

Similarly, the uncertain annealing events are:

$$\mathcal{U}_{\mu}(\alpha_{\mu}) = \left\{\beta:\ \beta^T W_{\mu}\beta \le \alpha_{\mu}^2\right\} \tag{8.55}$$

W_{λ} and W_{μ} are both real symmetric positive definite matrices.

Damage and annealing are both energetically controlled, and failure occurs when the cumulative energy level exceeds $E_{\rm cr}$. Let E_{λ} and E_{μ} denote the energy of damage and annealing, respectively, deposited in the system during a single load cycle. To keep track of the opposite effect of damage and annealing we treat E_{λ} as positive and E_{μ} as negative.

Due to the uncertainty in the input, $u(t)$, these energies each vary on an interval, whose limits are determined by the same method used to arrive at eq.(8.31):

$$0 \leq E_\lambda \leq c\alpha_\lambda^2 \nu_{\lambda,\max} \tag{8.56}$$

$$-c\alpha_\mu^2 \nu_{\mu,\max} \leq E_\mu \leq 0 \tag{8.57}$$

where:

$$\nu_{\lambda,\max} = \max \text{ eig} \left[W_\lambda^{-1/2} Z W_\lambda^{-1/2} \right] \tag{8.58}$$

$$\nu_{\mu,\max} = \max \text{ eig} \left[W_\mu^{-1/2} Z W_\mu^{-1/2} \right] \tag{8.59}$$

The system fails when the cumulative damage energy exceeds E_{cr}. Failure can be reached after ℓ damage cycles and m annealing cycles if the uncertainty parameters satisfy:

$$-cm\alpha_\mu^2 \nu_{\mu,\max} + c\ell\alpha_\lambda^2 \nu_{\lambda,\max} \geq E_{\text{cr}} \tag{8.60}$$

For fixed uncertainty in the annealing process, the critical value of the damage uncertainty α_λ, for damage state (ℓ, m), is obtained from (8.60) as:

$$\alpha_{\lambda,\ell m} = \sqrt{\frac{E_{\text{cr}} + cm\alpha_\mu^2 \nu_{\mu,\max}}{c\ell\nu_{\lambda,\max}}} \tag{8.61}$$

Likewise, when the damage uncertainty is fixed, the critical uncertainty in the annealing process is:

$$\alpha_{\mu,\ell m} = \sqrt{\frac{c\ell\alpha_\lambda^2 \nu_{\lambda,\max} - E_{\text{cr}}}{cm\nu_{\mu,\max}}} \tag{8.62}$$

Finally, to get a global measure of reliability we treat α_λ and α_μ as a single uncertainty parameter, α, whose critical value becomes:

$$\alpha_{\ell m} = \sqrt{\frac{E_{\text{cr}}}{c(\ell\nu_{\lambda,\max} - m\nu_{\mu,\max})}} \tag{8.63}$$

For each of these three expressions for robustness, eqs.(8.61)–(8.63), we can formulate an average robust reliability for lifetime Θ, randomized on the distribution $P_{\ell m}(\Theta)$. Consider first the damage reliability, with fixed uncertainty in the annealing process. Failure can occur after one or more damage events, with no constraint on the number of annealings which have taken place. We randomize on $\ell > 0$ and $m \geq 0$ to obtain:

$$\widehat{\alpha}_\lambda(\alpha_\mu) = \frac{\displaystyle\sum_{\ell\geq 1}\sum_{m\geq 0} \alpha_{\lambda,\ell m} P_{\ell m}(\Theta)}{\displaystyle\sum_{\ell\geq 1}\sum_{m\geq 0} P_{\ell m}(\Theta)} \tag{8.64}$$

Using eq.(8.62) we randomize the robust reliability of the annealing process with fixed damage uncertainty. The numerator in eq.(8.62) can be negative, so let $\ell_{\min}$ be the least value of ℓ for which $\alpha_{\mu,\ell m}$ takes a real value. Then the randomized robust reliability is:

$$\widehat{\alpha}_{\mu}(\alpha_{\lambda}) = \frac{\sum\limits_{\ell \geq \ell_{\min}} \sum\limits_{m \geq 0} \alpha_{\mu,\ell m} P_{\ell m}(\Theta)}{\sum\limits_{\ell \geq \ell_{\min}} \sum\limits_{m \geq 0} P_{\ell m}(\Theta)} \tag{8.65}$$

Finally, an expression for the overall randomized robust reliability is obtained with eq.(8.63). We must sum over all values of ℓ and m for which the denominator in eq.(8.63) is positive:

$$\widehat{\alpha} = \frac{\sum\limits_{\ell,m} \alpha_{\ell m} P_{\ell m}(\Theta)}{\sum\limits_{\ell,m} P_{\ell m}(\Theta)} \tag{8.66}$$

8.5 Problems

1. *Poisson probability and failure.* The solid propellant of a rocket is cast as a cylinder of length L and radius R, with a central bore of radius r. Cracks on the inner surface cause accelerated burning of the fuel. The presence of even one crack makes the unit unsafe. The procedure by which the central bore is produced results in λ cracks per unit area, on the average. (a) From a large sample of such items, what fraction are unusable? What assumptions have you made? (b) Given a single item chosen randomly from a large batch, what is the probability that it is usable? (c) From a large sample of items, all items having two or more cracks on the inner surface have been removed. What is the probability that an item chosen from the remaining population will be usable?

2. ‡ *Modification of problem 1.* We now consider cracks in the volume of the propellant, and not only on the inner surface of the bore. The average crack density is μ per cm^3. Furthermore, deep cracks are less dangerous than cracks near the bore surface, since combustion proceeds from the bore surface outwards, and thus meets deep cracks later than shallow ones. We have probabilistic information about the failure probability of the propellant, as a function of crack depth. Divide the fuel cylinder into M concentric shells with radii $r = r_0 < r_1 < \cdots < r_{M-1} < r_M = R$. The probability of failure given one or more cracks in the ith bin, (from r_{i-1} to r_i), is p_i. What is the overall probability of failure of the propellant? What assumptions have you made?

3. *Uncertain loads and cumulative damage.* Consider the damped oscillator of eq.(8.21) driven by inputs from a cumulative energy-bound convex model:

$$\mathcal{U}(\alpha) = \left\{ u(t) : \int_0^T [u(t) - \overline{u}(t)]^2 \, dt \leq \alpha^2 \right\} \tag{8.67}$$

The inputs occur as isolated random events distributed in time as a Poisson process. Each load cycle is of duration T and results in damage, δ, to the system which is determined by the maximum deflection during the cycle:

$$\delta = \eta \left[\max_{0 \leq t \leq T} x(t) \right]^{\nu} \tag{8.68}$$

η and ν are known constants. The system fails when the cumulative damage exceeds the critical value Δ_{cr}. (a) Derive an expression for the maximum damage, δ_{max}, possible in a single load cycle. (b) Derive an expression for α_n: the robust reliability for failure in n load cycles. (c) The system is considered to fail if the cumulative damage exceeds the critical value Δ_{cr} in a duration Θ. Randomize α_n on the Poisson distribution to obtain an average robust reliability.

4. ‡ *Uncertain loads.* Consider the damped oscillator of eq.(8.21) driven by inputs from the convex model of eq.(8.26). The inputs occur as isolated random events distributed in time as a Poisson process. (a) For fixed uncertainty, α, derive an expression for the least number, N_{cr}, of load cycles to failure. (b) Let $P_f(\Theta)$ be the probability of failure in time Θ, and let $P_n(\Theta)$ be the Poisson distribution. Prove that:

$$P_f(\Theta) \leq \sum_{n=N_{cr}}^{\infty} P_n(\Theta) \tag{8.69}$$

(c) Use this relation to derive an approximate robust reliability of the system.

5. *Shell imperfections.* Consider a thin-walled shell with band-limited imperfection profiles $\eta(r) = \beta^T \gamma(r)$ where $\gamma(r)$ is a known vector of trigonometric functions and β is an vector of uncertain Fourier coefficients constrained to an ellipsoid:

$$\mathcal{B}(\alpha) = \{ \beta : \beta^T W \beta \leq \alpha^2 \} \tag{8.70}$$

The spatial distribution of imperfections is described by the Poisson model, and the shell fails if the total imperfection amplitude exceeds Δ_{cr}. Develop a hybrid robust-probabilistic measure of the reliability of the shell.

6. *Modification of problem 5.* Replace the failure criterion by the following. The shell will fail with n imperfections if:

$$\int \left(\sum_{i=1}^{n} \eta_i(r) \right)^2 \mathrm{d}r \geq \Delta_{\mathrm{cr}}^2 \tag{8.71}$$

Develop a hybrid robust-probabilistic measure of reliability.

Chapter 9

Last But Not Final

We begin this chapter with a brief recapitulation of robust reliability, after which we address several remaining issues whose nature requires a somewhat more speculative approach than that adopted in previous chapters.

9.1 Recapitulation of Robust Reliability

We can rationally *rely* on something to the extent that our confidence is re-enforced by experience. In assessing reliability, we accumulate and evaluate information to dispell our uncertainty concerning the processes involved. This is the ordinary English meaning of reliability, and it underlies the technical meaning of reliability of engineering systems.

We evaluate the reliability of a mechanical system for various purposes: to determine the degree of safety of the system, or to estimate its lifetime, or to improve its design, or to enhance its quality. In order to exploit the quantitative tools of engineering analysis, we need a quantitative theory of reliability. That is, the intuitive lexical concept of reliability must be translated into a mathematical theory. This can be done in different ways. The classical theory of reliability is based on mathematical probability, and is exceedingly useful in those situations where we have sufficient information to verify the assumptions involved. In this book we have developed a theory of reliability based on convex-set models of uncertainty. Other reliability theories are also conceivable. In addition, one can combine these theories in various ways, such as the hybrid robust-probabilistic analysis discussed in chapter 8.

Robust reliability, based on convex models of uncertainty, measures reliability as the greatest amount of uncertainty which is consistent with no-failure of the system. A system is considered to be reliable if it is robust with respect to uncertainties in its constitution and operating environment. Conversely, if the system is fragile with respect to these uncertainties, if small deviations from nominal conditions can lead to failure, then the system is not reliable.

The uncertainty parameters of the convex models measure the robust reliability of the system.

The usual application of reliability analysis is in performance-evaluation of physical systems. However, the method of robust reliability can also be applied to mathematical models of mechanical systems. A model is reliable if the decisions which are based on the model are robust with respect to the uncertainties involved. Similarly, we can evaluate the reliability of algorithms for fault diagnosis or system identification in terms of their robustness to uncertainty.

Finally, a reliability theory is only as good as the information upon which it rests. A reliability theory should exploit all relevant verified information, but should treat speculative information and "reasonable assumptions" with caution. The convex models which underlie robust reliability are generally based on very fragmentary information. Convex models quantify uncertainty in terms of how events aggregate into clusters. Roughly speaking, this is usually less informative than a probabilistic model, which expresses uncertainty in terms of frequency of recurrence of events. When verified probabilistic information is available, it should be exploited. In designing and evaluating complex technological systems, especially under the severe time constraints which often dominate the industrial environment, prior information about uncertainties is often fragmentary. This is a major motivation for the use of convex rather than probabilistic models. However, some types of probabilistic information often is accessible, and can be incorporated into a hybrid probabilistic-robust reliability analysis.

In the remainder of this chapter we will discuss several open questions. In section 9.2 we examine the problem of calibrating the reliability: how reliable is reliable enough? In section 9.3 we consider the relation between reliability and the social acceptability of technological systems. Finally, in section 9.4 we discuss some managerial aspects of reliability.

9.2 Subjective Calibration of Robust Reliability

The robust reliability, $\widehat{\alpha}$, is measured as the greatest amount of uncertainty consistent with no-failure. $\widehat{\alpha}$ depends on the physical properties and design parameters of the system, and may vary in time as well. When $\widehat{\alpha}$ is large the system is robust to uncertainty, meaning that even substantial deviations from nominal conditions will not result in failure. On the other hand, when $\widehat{\alpha}$ is small the system is fragile and small variations entail the possibility of failure. The system is reliable when $\widehat{\alpha}$ is large; unreliable when $\widehat{\alpha}$ is small.

But how large is large enough?

There is no unique answer to this question. We will discuss two approaches, one based on evaluating the reliability in terms of the severity of failure-consequences, the other based on the magnitude of the information-gap which underlies the uncertainty.

9.2.1 Calibration by Consequence Severity

There are two aspects to the question of how large a value of $\widehat{\alpha}$ is required. First we can ask: what is the relation between the magnitude of $\widehat{\alpha}$ and the safety or acceptability of the system? How do we interpret the value of $\widehat{\alpha}$ in terms of human experience and expectations? Once one establishes the connection between the parameter $\widehat{\alpha}$ and subjective performance-attributes such as safety and acceptability, the second question to ask is: how safe is safe enough? Both of these questions are exceedingly difficult since the attributes in question, safety and acceptability, remain qualitative.

Concerning the first question, it is clear that "more is better" in reliability: safety and acceptability increase with $\widehat{\alpha}$. This implies that $\widehat{\alpha}$ is a quantitative tool for determining that one design is ***better than*** another. To go farther than this and to establish ***how much better*** one design is than another, it will be helpful to consider the severity of the consequences of failure.

We will focus the discussion on a simple example. Consider a critical rivet holding together two flanges and subjected to uncertain dynamic loads. As usual, the reliability analysis is based on three components:

1. A mechanical model describing the response of the rivet to the loads. This model includes various design parameters of the rivet. We will consider the dependence of the reliability on the diameter, D, of the rivet.

2. A convex model $\mathcal{U}(\alpha)$ describing the uncertainty of the load profiles, where the uncertainty parameter α assesses the degree of variability of the load functions.

3. A failure criterion. We will define failure to occur when the tensile stress in the rivet equals or exceeds the yield-point stress:

$$\sigma \geq \sigma_{\text{yp}} \tag{9.1}$$

For any choice of the rivet diameter D, the robust reliability, $\widehat{\alpha}$, is the greatest value of the load-uncertainty parameter α for which failure just becomes possible. When we wish to emphasize that the reliability depends on the choice of the design parameter we will write $\widehat{\alpha}(D)$.

The reliability will improve as the rivet diameter increases, as in fig. 9.1, since the applied loads are distributed over a greater cross-sectional area resulting in lower tensile stress. Clearly, a large rivet is more acceptable than a small rivet, but by how much? This is related to the global issue we are considering: how large a value of $\widehat{\alpha}$ is large enough? The first question mentioned above is: what connection can we make between the value of $\widehat{\alpha}$

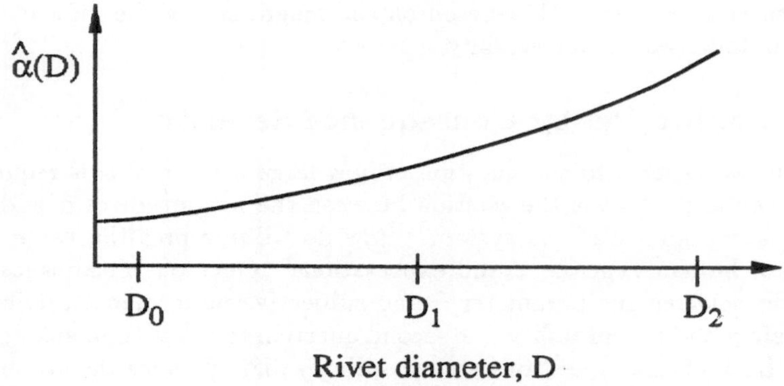

Figure 9.1: Variation of the reliability with rivet diameter.

and the safety and acceptability of the rivet? That is, how do we calibrate $\widehat{\alpha}$ in terms of subjective performance expectations? The second question is: how much safety should we require?

One way to calibrate the robustness $\widehat{\alpha}$ in terms of subjective performance expectations is to compare the reliability based on failure criteria of different severities. We will explain this by continuing the rivet example.

Criterion (9.1) is not the only option for defining failure of the rivet. One could choose a failure criterion which is either more or less conservative, corresponding to failure consequences which are either less or more severe: a *less* severe definition of failure results in a *more* conservative reliability analysis. For any choice of the failure criterion, the robust reliability, $\widehat{\alpha}$, is the greatest value of α for which failure just becomes possible. Let us consider three different criteria of failure, of increasing severity. We will then discuss the meaning of the corresponding robust measures of reliability, and their implication for interpreting the results in fig. 9.1 of the original reliability design analysis.

In all three criteria, failure is defined to occur when the tensile stress in the rivet exceeds a critical value:

$$\sigma \geq \sigma_{\rm cr} \tag{9.2}$$

Our three failure criteria are distinguished by the value of the critical stress:

1. *Low severity.* Failure is defined to occur when the stress reaches one-third of the yield-point stress:

$$\sigma_{\rm cr} = \frac{1}{3}\sigma_{\rm yp} \tag{9.3}$$

This is a conservative definition of failure, indicating that stresses are occurring which can, in the course of time, result in fatigue, micro-cracking, corrosion and other mechanisms of gradual damage evolution.

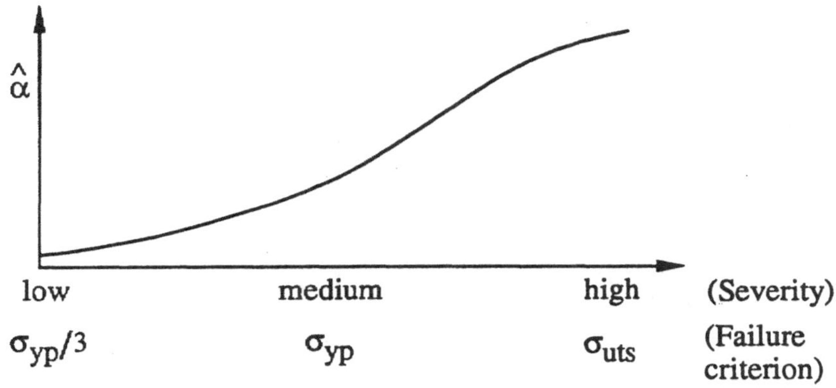

Figure 9.2: Robust reliability versus severity.

The consequences of failure in this definition are mild, indicating the need for maintenance or inspection.

2. *Medium severity.* Failure is defined to occur at the yield point:

$$\sigma_{cr} = \sigma_{yp} \tag{9.4}$$

The consequences of failure with this definition are more severe: large plastic deflection is possible under such large loads. This is the failure criterion of the original analysis, relation (9.1).

3. *High severity.* Failure occurs at the ultimate tensile strength of the material, and results in rupture of the rivet:

$$\sigma_{cr} = \sigma_{uts} \tag{9.5}$$

This is the least conservative failure criterion, since failure is defined as severe catastrophe.

Each one of these failure criteria generates a corresponding robust reliability, which we denote $\widehat{\alpha}_{low}$, $\widehat{\alpha}_{med}$ and $\widehat{\alpha}_{hi}$ for criteria (9.3), (9.4) and (9.5) respectively. The most conservative failure-criterion, eq.(9.3), generates the lowest measure of reliability, since "failure" occurs most easily. The high-severity criterion is associated with the greatest value of $\widehat{\alpha}$: catastrophe will occur only after large load deviations. Consequently, the magnitudes of these reliabilities will be ranked as:

$$\widehat{\alpha}_{low} < \widehat{\alpha}_{med} < \widehat{\alpha}_{hi} \tag{9.6}$$

We can in fact imagine a continuum of failure criteria from low to high severity, with a continuum of robustness, increasing as in fig. 9.2.

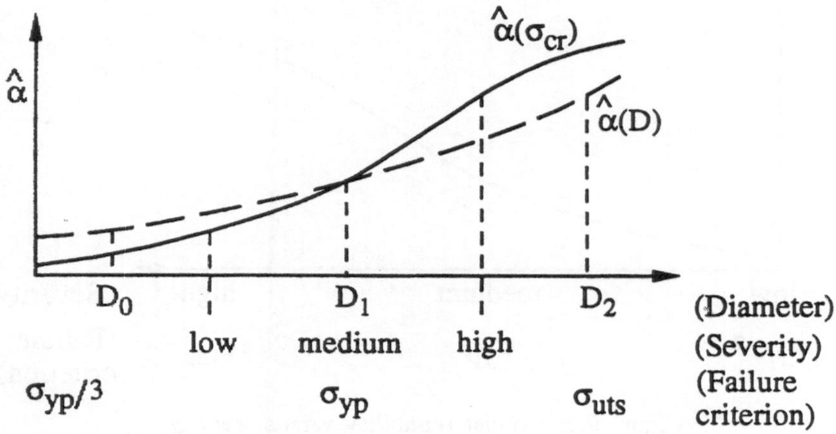

Figure 9.3: Overlay of figs. 9.1 and 9.2, giving an indication of the subjective advantage of different designs in terms of the severity of the failure consequences.

Before proceeding, we need to point out a common pitfall. We have said that large $\widehat{\alpha}$ is better than small $\widehat{\alpha}$, but in fig. 9.2 the reverse is the case: the unacceptable high-severity situation has the greatest value of $\widehat{\alpha}$. In fig. 9.2 we are comparing the robustnesses for *different failure criteria.* The assertion that "big is better" for $\widehat{\alpha}$ holds when comparing different system designs with *identical failure criteria.*

Now let us overlay figs. 9.1 and 9.2, as shown in 9.3. The solid line shows the variation of robustness versus consequence-severity, and the dashed line shows the variation with rivet diameter. The solid line is calculated for constant diameter D_1, and the dashed line is based on the medium-severity failure criterion, eq. (9.4). These two curves intersect at $D = D_1$ and $\sigma_{cr} = \sigma_{yp}$.

Rivet-diameter D_2 is more reliable than D_1. Fig. 9.3 gives us some indication of how much more reliable, in terms of subjective performance expectations. With diameter D_2, the rivet can tolerate, without experiencing medium-severity failure, an amount of load-uncertainty corresponding to the catastrophic failure criterion and diameter D_1. Diameter D_2 is subjectively much more reliable than D_1. Going in the reverse direction and comparing D_1 against D_0, we see that D_0 can tolerate, without medium-severity failure, an uncertainty consistent with low-severity failure and diameter D_1. In other words, D_0 is much less reliable, and hence much less acceptable, than D_1.

So how large an $\widehat{\alpha}$ is large enough? We must now consider the second question posed on p.207: how much safety do we require? This is not an engineering judgement, but rather a matter of personal and social perference. The engineering designer may be called upon to fulfil a variety of functions relating to the determination of acceptable design standards. He may be asked to identify available design options and to analyze their reliability. He may need to assess social and economic impacts of the various options. Finally, the engineer can and should make his own subjective safety assessment as an individual. We will have more to say about this in section 9.3. But ultimately, the subjective evaluation of safety and acceptability remains precisely that: subjective.

Let us conclude this discussion with a comparison between probabilistic and robust reliability. The usual probabilistic measure of reliability is the probability of no-failure, $P_{\rm nf}$. This probability is normalized to a maximum value of unity. Consequently, one has an intuitive understanding of the meaning of values near zero or near unity, even without reference to any particular system. Probability of no-failure near zero is clearly unacceptable, while values near unity are highly desirable. The uncertainty parameter of a convex model, α, is non-negative and unbounded, and we understand that uncertainty increases with α. Similarly the robust reliability index $\widehat{\alpha}$ can take any non-negative value, indicating the amount of tolerable uncertainty. Also, reliability increases with $\widehat{\alpha}$. Finally, using an analysis like that summarized in fig. 9.3 we can subjectively "calibrate" our understanding of the numerical value of $\widehat{\alpha}$. We thus can acquire an intuitive interpretation of any particular numerical value of $\widehat{\alpha}$.

However, in both the probabilistic and the robust theories, this intuitive feeling for the meaning of either $P_{\rm nf}$ or $\widehat{\alpha}$ is not as fundamental as it seems at first, since it does not indicate what values of either $P_{\rm nf}$ or $\widehat{\alpha}$ are *acceptable.* We may accept a probability of catastrophic failure in automobiles of 0.0001 per year, while in the nuclear or air travel industries we may require greater probabilistic reliability. In other words, the intuitive interpretation of either $P_{\rm nf}$ or $\widehat{\alpha}$ does not determine a criterion for acceptability. In both cases, only reference to considerations outside the reliability theory leads to acceptance or rejection of a given level of reliability. In this respect, probabilistic and robust reliabilities are identical: we must consult personal or public experience and preference in order to determine the design-goal values of the reliability, whether $P_{\rm nf}$ or $\widehat{\alpha}$.

9.2.2 Calibration by the Information Gap

We have shown in the previous section how to develop a subjective calibration of robust reliability by establishing a connection between values of $\widehat{\alpha}$ for different designs, and values of $\widehat{\alpha}$ for failure criteria of different severities. This is calibration in terms of consequence-severity. Now we perform a similar analysis, in which the subjective ranking of consequences is replaced by a subjective ranking of the amount of initial information. Instead of comparing different failure criteria, we compare different families of convex models. We will assess the robust reliability of various designs in terms of a subjective assessment of the initial information gap. This provides a different subjective

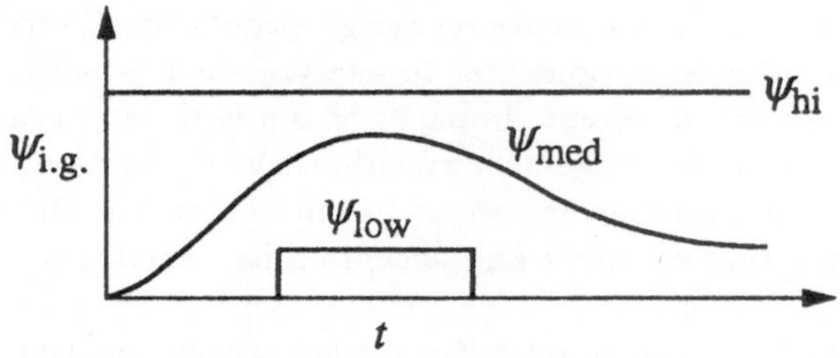

Figure 9.4: Envelope functions for low- medium- and high-information-gap convex models.

answer to the question on p.207: how much more reliable is one design than another?

We continue with the rivet example discussed in section 9.2.1, but now we concentrate on the convex uncertainty models of the unknown scalar load profile $u(t)$. We will consider three envelope-bound convex models, which we rank as low, medium and high, according to the extent of their inherent uncertainty. All three convex models are of the form:

$$\mathcal{U}_{\text{i.g.}}(\alpha) = \{u(t) : \ |u(t)| \leq \alpha\psi_{\text{i.g.}}(t)\}, \quad \text{i.g.} = \text{low, med, hi} \tag{9.7}$$

The envelope functions $\psi_{\text{i.g.}}$ for the low- medium- and high-information-gap models are non-negative functions ranked as:

$$\psi_{\text{low}}(t) \leq \psi_{\text{med}}(t) \leq \psi_{\text{hi}}(t) \tag{9.8}$$

for all values of t. Typical envelope functions are shown in fig. 9.4. For the same value of uncertainty parameter α, envelope ψ_{hi} implies the least constraint on the input and hence the greatest lack of prior information about the inputs which can occur, while ψ_{low} represents the tightest constraint on the input and the lowest information gap. The uncertainty parameters of all three convex models have the same units: those of the load function $u(t)$.

The uncertainty model $\mathcal{U}_{\text{low}}(\alpha)$ establishes a very "tight" definition of the input uncertainty, implying a small disparity between what *is* known about the input and what *could be* known: this is a small information gap. The intermediate convex model, $\mathcal{U}_{\text{med}}(\alpha)$, is somewhat less specific and has a greater information-gap for the same value of the uncertainty parameter. Finally, the third model, $\mathcal{U}_{\text{hi}}(\alpha)$, is the least specific, based on the loosest constraint on the inputs.

Let us suppose that the axial vibration of the rivet is linearly elastic. Consequently, for zero initial deflection and velocity of the rivet, the normal stress in the rivet in response to a load function $u(t)$, is described by:

$$\sigma_u(t) = \int_0^t u(\tau)z(t-\tau)\,\mathrm{d}\tau \tag{9.9}$$

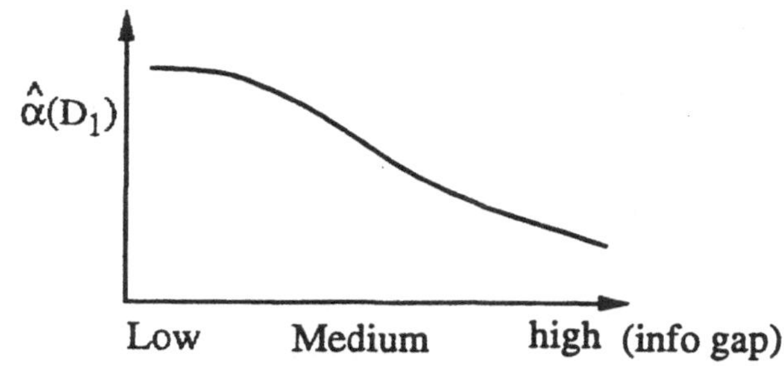

Figure 9.5: Robust reliability versus information gap.

where $z(t)$ is a known function depending on the properties of the rivet, in particular on the diameter D. The stress will decrease with increasing diameter, for the same load.

The rivet fails if the axial stress exceeds a critical value $\sigma_{\rm cr}$, as in eq.(9.2). The greatest axial stress in the rivet, for any load profile in a particular convex model $\mathcal{U}_{\rm i.g.}(\alpha)$, occurs when $u(\tau)$ follows the envelope, $\psi_{\rm i.g.}(\tau)$, switching between the upper and lower branches as $z(t-\tau)$ changes sign:

$$\widehat{\sigma}_{\rm i.g.}(t) = \max_{u\in\mathcal{U}_{\rm i.g.}(\alpha)} \sigma_u(t) \tag{9.10}$$

$$= \alpha \int_0^t \psi_{\rm i.g.}(\tau)|z(t-\tau)|\,{\rm d}\tau \tag{9.11}$$

for i.g. = low, med, hi. In light of the ranking of the envelope functions in relation (9.8), the maximum stresses for each of the three convex models are similarly ranked, for fixed rivet diameter:

$$\widehat{\sigma}_{\rm low}(t) \le \widehat{\sigma}_{\rm med}(t) \le \widehat{\sigma}_{\rm hi}(t) \tag{9.12}$$

The robust reliability for each convex model can be expressed by equating $\widehat{\sigma}_{\rm i.g.}$ to the critical stress and then solving for α:

$$\widehat{\sigma}_{\rm i.g.} = \sigma_{\rm cr} \quad \Longrightarrow \quad \widehat{\alpha}_{\rm i.g.} = \frac{\sigma_{\rm cr}}{\int_0^t \psi_{\rm i.g.}(\tau)|z(t-\tau)|\,{\rm d}\tau} \tag{9.13}$$

From relations (9.12) and (9.13) one sees that the reliabilities for the three convex models are ranked in reverse order:

$$\widehat{\alpha}_{\rm low}(t) \ge \widehat{\alpha}_{\rm med}(t) \ge \widehat{\alpha}_{\rm hi}(t) \tag{9.14}$$

This ranking of course is for a given value of the rivet diameter. We can imagine a continuum of convex models, whose information-gaps vary from low to high. The reliabilities corresponding to these uncertainty models will also decrease continuously, as shown in fig. 9.5.

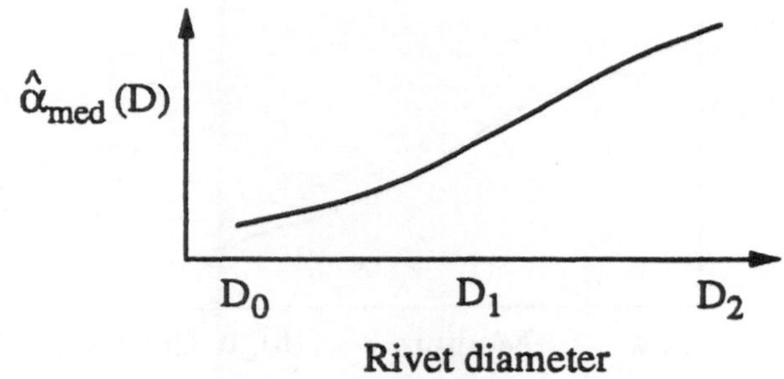

Figure 9.6: Variation of the reliability with rivet diameter; medium information gap.

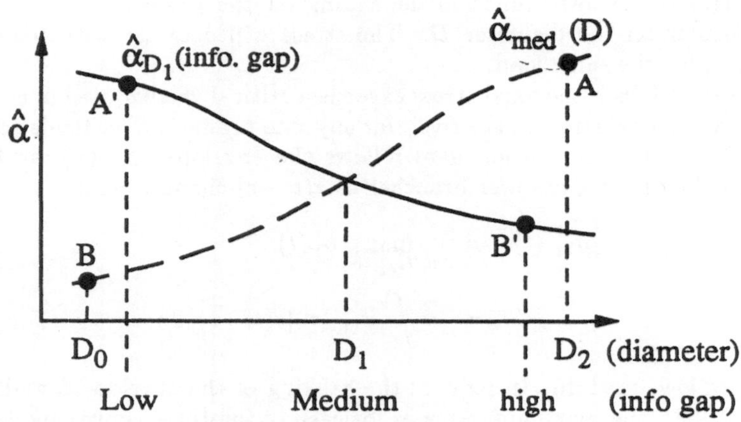

Figure 9.7: Overlay of figs. 9.5 and 9.6, giving an indication of the subjective advantage of different designs in terms of the information gap inherent in the convex model.

But of course if we consider only a particular level of uncertainty, for instance $\mathcal{U}_{\rm med}(\alpha)$, and vary the rivet diameter, then the reliability will vary like fig. 9.6. The basic question is: how much does the *subjective* reliability vary over this range of designs?

To assess this in terms of the initial information gap, we overlay figs. 9.5 and 9.6, as in fig. 9.7. The rising dashed line is the robust reliability as a function of rivet diameter, evaluated with $\mathcal{U}_{\rm med}(\alpha)$, the medium-information-gap model of uncertainty. The decreasing solid line is the reliability of rivet size D_1, as a function of the subjective magnitude of the information gap. These curves cross exactly at diameter D_1 and medium info-gap size.

Diameter D_2 is substantially more reliable than size D_1, as we see by comparing points A and A' in fig. 9.7. Diameter D_2 and the medium-gap model can tolerate, without experiencing failure, an amount of load-uncertainty corresponding to the low-info-gap uncertainty model and diameter D_1. In other words, increasing the diameter from D_1 to D_2 enhances the reliability in a way which is in some sense equivalent to reducing the load-uncertainty from the medium- to the low-information-gap level. The difference between design values D_1 and D_2 is in this sense subjectively equivalent to the difference between the medium- and low-gap uncertainty models. Conversely, diameter D_0 is much less reliable than D_1, as seen from points B and B'. Diameter D_0 and the medium-gap uncertainty model can tolerate, without failure, only as much uncertainty as D_1 and the high-gap model.

The qualitative calibrations of $\widehat{\alpha}$ discussed in this section and the previous one provide alternative assessments of the degree of reliability. These assessments, being subjective, need not be entirely consistent with one another. They represent different aspects of the difficult problem of correlating numbers with anticipations and expectations.

9.3 Reliability and Social Acceptability

The reliability of technological constructions is one of the oldest concerns of civilization. Forty centuries ago Hammurabi decreed the following severe penalties for structural failure:

> If a mason has built a house for a man, but (if) he has not consolidated his work and (if) the house he has built falls down causing the death of the owner of the house, this mason shall be killed. If he causes the death of the child of the owner of this house, the mason's child shall be killed. (Articles 229–230)

Reliability, as an aspect of human affairs and an issue in society, goes beyond the scope of quantitative analyses like the one developed in this book. Social priorities, subjective evaluation of risks, economic constraints, environmental impact, hidden costs and delayed effects, unpredictable even unquantifiable human behavior: all are relevant to objective as well as perceived reliability.

When the engineer evaluates his invention or design in the context of social implications and issues, he faces a confusing kaleidoscope of concerns. The engineer may view his technological system as residing at the hub of a merry-go-round of supra-technological complexities (fig. 9.8). His formal professional competence is constrained to the technical system itself, but at least part of his effort must be directed outside the immediate domain of technology. Cognizance of social expectations can profoundly influence his professional decisions. Just ask any mason from Hammurabi's time!

So, must the responsible engineer, concerned about the social acceptability of his product as well as about his own well-being, be an economist, ecologist, psychologist, moralist and lawyer, all in one? No, but of course also yes. The engineer has multi-faceted responsibilities which transcend

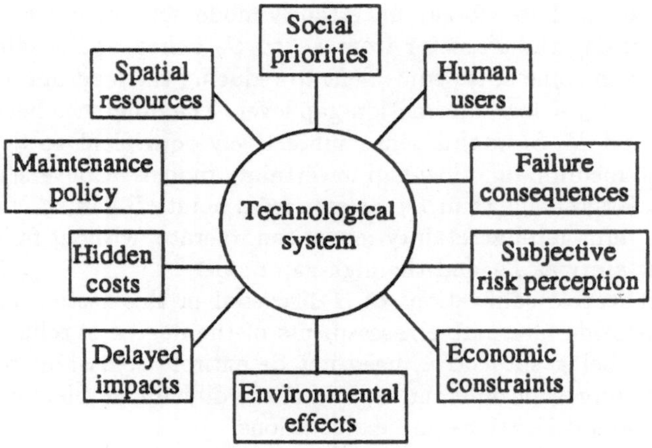

Figure 9.8: Social issues in technological reliability.

his purely technical function. One result is that many engineering ventures involve teamwork often on a grand scale, between specialists of many disciplines, scientific, technical and social.

But as issues get more complex and interwoven, as suggested in fig. 9.8, the advantages of the expert begin to fade, and not all responsibility can be relegated to topical specialists. Each member of the team must contribute to achieving social acceptability for the system being developed, designed, constructed, tested, marketed, serviced or repaired. Each member of the team must identify the uncertainties which accompany the mission, though often these are not quantifiable with mathematical models. One must identify the information gap between what is known and what could be known. It is a highly speculative, subjective and tenuous task, but one must attempt to enhance the robustness of the mission to these uncertainties. This *speculative reliability* is far more difficult to achieve, or even to detect once it has been achieved, than the quantitative technical reliability we have studied in this book. But nonetheless, in applying the technical concepts of reliability, the engineer should recognize his involvement in the broader goal of achieving social acceptability for his product.

9.4 Robustness as a Managerial Strategy

Robustness to uncertainty has been recognized as a guiding principle for reliable management with deficient information. Laufer identifies nine principles for "managing projects in an era of uncertainty". He writes that one central principle in project management is that, "in dynamic conditions, influencing the future is not just making decisions early. It is more about the ability to **reduce uncertainty** and to **minimize the impact of surprises.**" (emphasis in the original) [57]. This is a direct statement of the need for robustness to uncertainty.

One implication is that technical product-reliability should be highlighted on the managerial agenda. Unanticipated technical failures can lead to the most unpleasant surprises. The O-ring failure during the Challenger launch of 1985 is a prime example. A single component was given minor attention by the engineers because of its technical simplicity and universality. Its failure however led to catastrophic termination of the mission when the entire space vehicle exploded shortly after launch. For the project manager, reducing uncertainty and minimizing the impact of surprises can mean, among other things, enhancing the robust reliability of components and systems. In complex integrated engineering projects, this must be a central concern of the managerial as well as of the technical staff.

Managers are often engineers, having begun their careers in a specific technical discipline. This can facilitate the manager's ability to communicate effectively with the engineering staff in coordinating the management of reliability. Engineers, on the other hand, may have limited managerial experience. Since communication is a two-way street, problems can develop in establishing effective overall control of system-wide reliability. In the training of engineers, both at the university and in industry, it is important to enhance the systems-consciousness of the engineer, to strengthen his awareness of the interconnections among physical sub-units and between temporal stages of the project. For example, the engineer as well as the manager must be aware that decisions made by the designer at the conceptual stage can have profound impact on the actions and options of the maintenance engineer much later in the life of the product. In short, robustness is a useful concept for the management of uncertainty in complex projects and, conversely, the integrative perspective of the project manager is at times relevant to specific technical decisions.

References

1. Abrahamson, G.R. and Goodier, J.N., 1962, Dynamic Plastic Flow Buckling of a Cylindrical Shell from Uniform Radial Impulse, 4th U.S. National Congress of Applied Mechanics, Berkeley, CA, pp 939–950.

2. J. Arbocz and J.G. Williams, 1977, Imperfection surveys on a 10 ft. diameter shell structure, *Amer. Inst. of Aeronautics and Astronautics Journal*, 15: 949–956.

3. Z. Artstein, 1974, On the calculus of closed set-values functions, *Indiana University Mathematics Journal*, 24: 433-441.

4. Z. Artstein and J.C. Hansen, 1985, Convexification in limit laws of random sets in Banach spaces, *Annals of probability*, 13: 307–309.

5. R.J. Aumann, 1965, Integrals of set-valued functions, *Journal of Mathematical Analysis and Applications*, 12:1–12.

6. Y. Ben-Haim, 1985, *The Assay of Spatially Random Material*, Kluwer Academic Publishers, Dordrecht, Holland.

7. Y. Ben-Haim, 1990, Optimizing multi-hypothesis diagnosis of control-actuator failures in linear systems, *AIAA Journal of Guidance, Control and Dynamics*, 13: 744-750.

8. Y. Ben-Haim, 1990, Detecting unknown lateral forces on a bar by vibration measurement, *Journal of Sound and Vibration*, 140: 13–29.

9. Y. Ben-Haim, 1992, Convex models for optimizing diagnosis of uncertain slender obstacles on surfaces, *Journal of Sound and Vibration*, 152: 327–341.

10. Y. Ben-Haim, 1992, Adaptive Diagnosis of Faults in Elastic Structures by Static Displacement Measurement: The Method of Selective Sensitivity. *Mechanical Systems and Signal Processing*, 6: 85–96.

11. Y. Ben-Haim, 1993, Convex models of uncertainty in radial pulse buckling of shells, *ASME Journal of Applied Mechanics*, 60:683–688.

12. Y. Ben-Haim, 1993, Failure of an axially compressed beam with uncertain initial deflection of bounded strain energy, *International Journal of Engineering Science*, 31: 989–1001.

13. Y Ben-Haim 1993, Identification of certain polynomial nonlinear structures by adaptive selectively-sensitive excitation. *ASME Journal of Vibration and Acoustics*, 115:246–255.

14. Y. Ben-Haim, 1994, A non-probabilistic concept of reliability, *Structural Safety*, 14:227–245.

15. Y. Ben-Haim, 1994, Fatigue lifetime with load uncertainty represented by convex model, *ASCE J. Engineering Mechanics*, 120: 445–462.

16. Y. Ben-Haim, 1994, Convex models of uncertainty: Applications and Implications, *Erkenntnis: An International Journal of Analytic Philosophy*, 41:139–156.

17. Y. Ben-Haim, 1995, A Non-Probabilistic Measure of Reliability of Linear Systems Based on Expansion of Convex Models, *Structural Safety*, 17:91–109.

18. Y. Ben-Haim and E. Elias, 1987, Indirect measurement of surface temperature and heat flux: optimal design using convexity analysis, *International Journal of Heat and Mass Transfer*, 30: 1673–1683.

19. Y. Ben-Haim and I. Elishakoff, 1989, Non-Probabilistic models of uncertainty in the non-linear buckling of shells with general imperfections: Theoretical estimates of the knockdown factor, *ASME Journal of Applied Mechanics*, 56: 403–410.

20. Y. Ben-Haim and I. Elishakoff, 1990, *Convex Models of Uncertainty in Applied Mechanics*, Elsevier, Amsterdam.

21. Y. Ben-Haim and I. Elishakoff, 1991, Convex models of vehicle response to uncertain but bounded terrain. *ASME Journal of Applied Mechanics*, 58: 354–361.

22. Y. Ben-Haim and H. G. Natke, 1992, Diagnosis of Changes in Elastic Boundary Conditions in Beams by Adaptive Vibrational Testing. *Archive of Applied Mechanics*, 62: 210–221.

23. Y. Ben-Haim and U. Prells, 1993, Selective Sensitivity in the Frequency Domain, Part I: Theory. *Mechanical Systems and Signal Processing*, 7: 461–475.

24. Y. Ben-Haim and N. Shenhav, 1984, The measurement of spatially random material, *SIAM Journal of Applied Mathematics*, 44:1150–1163.

25. S. Braun, ed., 1986, *Mechanical Signature Analysis: Theory and Applications,* Academic Press.

26. L. Carroll 1895, *Pillow Problems,* Re-issued by Dover Press, New York, 1958.

27. C. Cempel, 1991, Pareto law of damage evolution in application to vibration-condition monitoring, *Bulletin of the Polish Academy of Sciences,* vol. 39, nr. 4, pp 573–588.

28. C. Cempel and H.G. Natke, 1993, Damage evolution and diagnosis in operating systems, in: *Safety Evaluation Based on System Identification Approaches Related to Time-Variant and Nonlinear Structures,* H.G. Natke, *et al,* eds., Vieweg Braunschweig, pp44-61.

29. S. Cogan, F. Ayer, Y. Ben-Haim and G. Lallement, 1995, Updating linear elastic models with modal selective sensitivity, *Inverse Problems in Engineering,* 2: 29–47.

30. J.L. Coolidge, 1990, *The Mathematics of Great Amateurs,* Clarendon Press, Oxford, 2nd edition.

31. B. De Finetti, 1969, Initial probabilities: A prerequisite for any valid induction, *Synthese,* 20: 2–16.

32. B. De Finetti, 1974, *Theory of Probability: A Critical Introductory Treatment,* vol. 1, trans. by A. Machi and A. Smith, John Wiley, London.

33. R.F. Drenick, 1968, Functional analysis of effects of earthquakes, *2nd Joint United States-Japan Seminar on Applied Stochastics,* Washington, D.C., Sept. 19–24.

34. R.F. Drenick, 1970, Model-free design of aseismic structures, *Journal of the Engineering Mechanics Division, Proceedings of the ASCE,* 96: 483–493.

35. D. Dubois and H. Prade, 1988, *Possibility Theory: An Approach to Computerized Processing of Uncertainty,* Plenum Press, New York.

36. I. Elishakoff, 1983, *Probabilistic Methods in the Theory of Structures,* Wiley, New York.

37. I. Elishakoff and Y. Ben-Haim, 1990, Dynamics of a thin shell under impact with limited deterministic information on its initial imperfections, *International Journal of Structural Safety,* 8: 103–112.

38. I. Elishakoff and L.P. Zhu, 1994, *Probabilistic and Convex Modelling of Acoustically Excited Structures,* Elsevier, Amsterdam.

39. P. Eykhoff, 1974, *System Identification: Parameter and State Estimation*, John Wiley, Chichester.

40. W. Feller, 1970, *An Introduction to Probability Theory and its Applications*, vol. 1, 3rd ed., John Wiley.

41. J.E. Fenstad, 1968, The structure of logical probabilities, *Synthese*, 18: 1–12.

42. M.I. Friswell and J.E. Mottershead, 1995, *Finite Element Model Updating in Structural Dynamics*, Kluwer Academic Publishers, Dordrecht.

43. B.R. Gaines, 1978, Fuzzy and probability uncertainty logics, *Information and Control*, 38: 154–169.

44. J. Galbraith, 1973, *Designing Complex Organizations*, Addison-Wesley Publ. Co.

45. P. Gärdenfors and N.-E. Sahlin, 1982, Unreliable probabilities, risk taking and decision making, *Synthese*, 53: 361–386.

46. C.F. Gauss, 1823, *Theory of the Combination of Observations Least Subject to Errors*, Trans. by G.W. Stewart, Society for Industrial and Applied Mathematics, Philadelphia, 1995.

47. J.D. Gibbons and S. Chakraborti, 1992, *Nonparametric Statistical Inference*, 3rd edition, Marcel Dekker, New York.

48. G.H. Hardy, J.E. Littlewood and G. Polya, 1934, *Inequalities*, Cambridge University Press, London.

49. W.H. Hartt and N.K. Lin, 1986, A proposed stress history for fatigue testing applicable to offshore structures, *International Journal of Fatigue*, 8: 91–93.

50. M. Hollander and D.A. Wolfe, 1972, *Nonparametric Statistical Methods*, Wiley, New York.

51. S.L. Hurley, 1991, Newcomb's problem, prisoner's dilemma, and collective action, *Synthese*, 86: 173–196.

52. P.J. Kelly and M.L. Weiss, 1970, *Geometry and Convexity: A Study in Mathematical Methods*, Wiley, New York.

53. S.W. Kirkpatrick and B.S. Holmes, 1989, Effect of initial imperfections on dynamic buckling of shells, *ASCE Journal of Engineering Mechanics*, 115: 1075–1093.

54. A.N. Kolmogorov, 1933, *Foundations of the Theory of Probability*, 2nd English ed. trans. by N. Morrison, based on the 1933 German monograph.

55. H.E. Kyburg, jr., 1990, Getting fancy with probability, *Synthese*, 90: 189–203.

56. P. Lancaster and M. Tismenetsky, 1985, *The Theory of Matrices*, 2nd ed., Academic Press.

57. A. Laufer, 1996, *Simultaneous Management: Managing Projects in an Era of Uncertainty and Accelerated Speed*, AMACOM, American Management Association, New York.

58. I. Levi, 1982, Ignorance, probability and rational choice, *Synthese*, 53: 387–417.

59. F.S. Li and S. Kyriakides, 1991, On the Propagation Pressure of Buckles in Cylindrical Confined Shells, ASME *Journal Applied Mechanics*, Vol. 57, pp 1091–1094.

60. Lindberg, H.E. and Florence, A.L., 1987, *Dynamic Pulse Buckling*, Martinus Nijhoff Publishers, Dordrecht, The Netherlands, Kluwer Academic Publishers, U.S. and Canada distributer, Hinghand, MA.

61. H.E. Lindberg, 1992, An evaluation of convex modelling for multimode dynamic buckling, *ASME Journal of Applied Mechanics*, 59: 929–936.

62. H.E. Lindberg, 1992, Convex models for uncertain imperfection control in multimode dynamic buckling, *ASME Journal of Applied Mechanics*, 59: 937–945.

63. L. Ljung, 1987, *System Identification: Theory for the User*, Prentice Hall, Englewood Cliffs.

64. J.G. March, 1994, *A Primer on Decision Making: How Decisions Happen*, The Free Press, A Division of Macmillan, Inc., New York.

65. V.A. Morozov, 1984, *Methods for Solving Incorrectly Posed Problems*, Z. Nashed translation editor, Springer-Verlag, New York.

66. J.E. Mottershead and C.D. Foster, 1991, On the treatment of ill-conditioning in spatial parameter estimation from measured vibration data, *Mechanical Systems and Signal Processing*, 5:139–154.

67. K. Murota and K. Ikeda, 1992, On random imperfections for structures of regular-polygonal symmetry, *SIAM J. Appl. Math.* 52:1780–1803.

68. H.G. Natke, D. Collmann, H. Zimmermann, 1974, Beitrag zur Korrektur des Rechenmodells eines elastomechanischen Systems anhand von Versuchsergebnissen, VDI-Berichte nr. 221, pp23–32.

69. H.G. Natke, 1979, Vergleich von Algorithmen für die Anpassung des Rechenmodells einer schwingungsfähigen elastomechanischen Struktur an Versuchswerte, *Zeitschrift für Angewandte Mathematik und Mechanik,* 59: 257–268.

70. H.G. Natke, 1991, Error localization within spatially finite-dimensional mathematical models: A review of methods and the application of regularization techniques, *Computational Mechanics,* 8: 153-160.

71. H.G. Natke, 1992, On regularization methods within system identification, in *Inverse Problems in Engineering Mechanics,* M. Tanaka and H.D. Bui, eds., IUTAM Symposium, Tokyo. Springer-Verlag, Berlin.

72. H.G. Natke and C. Cempel, 1994, Holistic modelling as a tool for the diagnosis of critical complex systems, IFAC Symposium on Fault Detection, Supervision and Safety for Technical Processes, Safeprocess '94, Helsinki, 13–16 June 1994, pp288–293.

73. H.G. Natke and T.T. Soong, 1993, Topological structural optimization under dynamic loads, in S. Herndandex and C.A. Brebbia, eds., *Optimization of Structural Systems and Applications,* Computational Mechanics Publications, Southampton, U.K. pp.67–78.

74. D.E. Newland, 1986, General linear theory of vehicle response to random road roughness, in I. Elishakoff and R.H. Lyons, eds., *Random Vibrations — Status and Recent Developments,* Elsevier Science Publishers, Amsterdam, pp.303–326.

75. J. Pearl, 1988, *Probabilistic Reasoning in Intelligent Systems: Networks of Plausible Inference,* Morgan Kaufman Publishers, San Mateo.

76. U. Prells and Y. Ben-Haim, 1993, Selective Sensitivity in the Frequency Domain, Part II: Applications. *Mechanical Systems and Signal Processing,* 7:551–574.

77. W.C. Reynolds and H.C. Perkins, 1977, *Engineering Thermodynamics,* 2nd ed., McGraw-Hill Book Co.

78. R.T. Rockafellar, 1970, *Convex Analysis,* Princeton University Press, Princeton, N.J.

79. R. T. Rockafellar and W. J.-B. Wets, 1991, Scenarios and policy aggregation in optimization under uncertainty, *Mathematics of Operations Research,* 16:119–147.

80. F.C. Schweppe, 1968, Recursive state estimation: Unknown but bounded errors and system inputs, *IEEE Transactions on Automatic Control,* AC-13: 22–28.

81. F.C. Schweppe, 1973, *Uncertain Dynamic Systems,* Prentice-Hall, Englewood Cliffs.

82. I.H. Shames, 1982, *Mechanics of Fluids,* 2nd ed., McGraw-Hill.

83. N. Shenhav and Y. Ben-Haim, 1984, A general method for optimal design of nondestructive assay systems, *Nuclear Science and Engineering,* 88: 173–183.

84. M. Shinozuka, 1970, Maximum structural response to seismic excitations, *Journal of the Engineering Mechanics Division, Proceedings of the ASCE,* 96: 729–738.

85. J. Skilling, 1985, Prior probabilities, *Synthese,* 63: 1–34.

86. K. Sobczyk and B.F. Spencer, 1992, *Random Fatigue: From Data to Theory,* Academic Press, Boston.

87. T. Söderström and P. Stoica, 1989, *System Identification,* Prentice Hall, Englewood Cliffs.

88. T.T. Soong, 1981, *Probabilistic Modelling in Science and Engineering,* Wiley, New York.

89. T.T. Soong, 1990, *Active Structural Control: Theory and Practice,* Wiley, New York.

90. T.T. Soong, G. Chen, Z. Wu, R.-H. Zhang and M. Grigoriu, 1993, Assessment of the 1991 NEHRP Provisions for Nonstructural Components and Recommended Revisions, Natl. Center for Earthquake Engg. Res., State Univ. of N. Y. at Buffalo, report NCEER-93-0003, March 1993.

91. S.M. Stigler, 1986, *The History of Statistics: The Measurement of Uncertainty Before 1900,* Belnap Press, Cambridge.

92. J. Stoer and C. Witzgall, 1970, *Convexity and Optimization in Finite Dimensions,* Springer Verlag, Berlin.

93. P. Suppes and M. Zanotti, 1977, On using random relations to generate upper and lower probabilities, *Synthese,* 36: 427–440.

94. G. Taguchi, E.A. Elsayed and T.C. Hsiang, 1989, *Quality Engineering in Production Systems,* McGraw-Hill, New York.

95. A. Talmor, Y. Leichter, Y. Ben-Haim and A. Kushlevsky, 1990, Adaptive assay of radioactive pulmonary aerosol with an external detector. *International Journal of Applied Radiation and Isotopes,* Part A, 41: 989–993.

96. J. Terney, 1991, Behind Monty Hall's doors: Puzzle, debate and answer, *New York Times*, 21 July 1991, section I, p1.

97. A.N. Tikhonov and V.Y. Arsenin, 1977, *Solutions of Ill-Posed Problems,* Fritz John translation editor, John Wiley and Sons, New York.

98. S.P. Timoshenko and J.M. Gere, 1963, *Theory of Elastic Stability,* 2nd ed., McGraw-Hill.

99. J. Venn, 1888, *The Logic of Chance,* Chelsea Publishing Co., New York, 4th edition, 1962, which is an unaltered reprint of the 3rd, 1888, edition.

100. Vinograd, I.M., ed., 1991, *Encyclopedia of Mathematics,* Kluwer Academic Publishers, Dordrecht, Holland.

101. U. Vulkan and Y. Ben-Haim, 1989, Global optimization in the adaptive borehole assay of Uranium, *International Journal of Applied Radiation and Isotopes*, Part E: *Nuclear Geophysics*, 3: 97–105.

102. M. Wakabayashi, 1986, *The Design of Earthquake-Resistant Buildings,* McGraw-Hill Book Co., New York.

103. M.M.R. Williams, 1974, *Random Processes in Nuclear Reactors,* Pergamon Press, Oxford.

104. H.S. Witsenhausen, 1968, A minimax control problem for sampled linear systems, *IEEE Transactions on Automatic Control,* AC-13: 5–21.

105. H.S. Witsenhausen, 1968, Sets of possible states of a linear system given perturbed observations, *IEEE Transactions on Automatic Control,* AC-13: 556–558.

106. J.T.P. Yao, 1985, *Safety and Reliability of Existing Structures,* Pitman, Boston.

107. R.H. Zhang and T.T. Soong, 1992, Seismic design of viscoelastic dampers for structural applications, *ASCE Journal of Structural Engineering,* 118:1375–1392.

Author Index

Subject Index

Springer-Verlag and the Environment

We at Springer-Verlag firmly believe that an international science publisher has a special obligation to the environment, and our corporate policies consistently reflect this conviction.

We also expect our business partners – paper mills, printers, packaging manufacturers, etc. – to commit themselves to using environmentally friendly materials and production processes.

The paper in this book is made from low- or no-chlorine pulp and is acid free, in conformance with international standards for paper permanency.